KRAFTWERKE
POWER STATIONS
IN BERLIN

Das Erbe der Elektropolis – The Electropolis Heritage

INHALT

CONTENTS

Kraftwerk Charlottenburg, Maschinenhalle, Detailaufnahme Fassade, um 1980
Charlottenburg Power Station, machine hall, detailed view of façade, photograph around 1980

HISTORISCHE KRAFTWERKE IN UMNUTZUNG
THE RE-UTILISATION OF HISTORIC POWER STATIONS

Denkmalverantwortung und Entwicklungspotenzial
Responsibility to Monuments and Development Potential

Hans Achim Grube
Herbert Strobel

Vor fast 120 Jahren, im Jahre 1884, begann in Berlin die Geschichte der Energieversorgung. Die Bewag und ihre Vorgängergesellschaften mit ihrem visionären Gründer Emil Rathenau wurden zu einem wesentlichen Taktgeber für die Industrialisierung Berlins. Der Stromverbrauch stieg stetig, phasenweise gleichsam explosionsartig. Immer neue Kraftwerke wurden notwendig und innerhalb kürzester Zeiträume errichtet. Noch heute sind so traditionelle Kraftwerksstandorte wie die Heizkraftwerke Charlottenburg, Moabit und Klingenberg erkennbare Zeugnisse dieser Entwicklung.
Durch die Liberalisierung des bundesdeutschen Strommarktes Mitte der 1990er Jahre erfolgte eine dramatische Veränderung der Unternehmensstrukturen des ehemaligen Monopolunternehmens. Strom wurde zur ortsungebundenen Handelsware, er wurde nicht mehr notwendigerweise vor Ort produziert, sondern kostengünstig eingekauft. Lediglich fernwärmegesteuerte Heizkraftwerke behielten den lokalen Bezug.

Somit sind einige Kraftwerke des Berliner Energieversorgers Bewag durch die technische Weiterentwicklung teilweise funktionslos geworden. Trotz des langjährigen baulichen Instandhaltungsprogramms der Bewag kann der weitere Erhalt dieser Gebäudekomplexe nur durch entsprechende Nachnutzungskonzepte erreicht werden, da ansonsten ein schrittweiser Verfall durch Leerstand und Vandalismus droht.
Die Bewag hat durch eine systematische Bestandserfassung der vorhandenen Bausubstanz die Grundlagen für eine immobilienwirtschaftliche und funktionsorientierte Bewertung der einzelnen Gebäude geschaffen. Die Denkmalklassifizierung der einzelnen Bauten wird vom Eigentümer und der Denkmalpflege als Chance begriffen.
Das immobilienwirtschaftliche Ziel der Bewag besteht seit 1997 darin, durch gezielte konzeptionelle Umnutzungsplanungen und Marketingmaßnahmen einen Markt für denkmalgeschützte Sondergebäude der

Almost 120 years ago, in 1884, the history of energy provision began in Berlin. The Bewag (Berliner Elektrizitätswerke AG) and the companies that preceded it, along with their visionary founder Emil Rathenau, became an instrumental part in the industrialisation of Berlin. Use of electricity increased constantly, at times, almost exponentially. More and more new power stations were needed, and constructed with as little delay as possible. Even today such traditional power station sites as the thermal power plants of Charlottenburg, Moabit and Klingenberg are obvious signs of this development.
The liberalisation of the Federal German power market in the mid-1990s brought about a dramatic change in the structure of what used to be a monopoly. Power became a non-localised form of merchandise; it no longer had to be produced where it was needed, but could be purchased at the best price. Only thermal power plants that were fed by long-distance thermal energy retained a local connection.

Accordingly, as a result of further technical developments some of the power stations operated by the Berlin energy-supplier Bewag became – in part, at least – functionless. Despite the fact that the Bewag has, for many years, kept up a programme of building-maintenance, the continued existence of these building complexes can be guaranteed only by suitable strategies for re-utilisation, as otherwise, they are in danger of gradual collapse by standing empty or because of vandalism. By undertaking a detailed assessment of the state of existing buildings, the Bewag has established the basis for evaluating individual buildings in terms of their land and function. The classification of individual buildings as monuments to be preserved is seen as an opportunity by both the owners and those responsible for monument preservation.

Kraftwerk Rummelsburg, Ansicht von der Rummelsburger Landstraße, Aufnahme um 1992
Rummelsburg Power Station, view from Rummelsburger Landstraße, photograph around 1992

Elektrizitätswirtschaft zu schaffen und die Gebäude an Eigennutzer zu veräußern oder zu vermieten. Die Bewag, das Landesdenkmalamt und freie Architekten erarbeiteten gemeinsam als neue Form des Denkmalmanagements Erhaltungskonzepte für diese technischen Baudenkmale. Die schützenswerten Erhaltungsbereiche werden dokumentiert und es werden Spielräume zur baulichen und funktionalen Weiterentwicklung, Nach- und Neunutzung entwickelt.

Dieser konstruktive Prozess wurde mit der Publikation „Elektropolis Berlin" (Hrsg.: Bewag, Berlin 2000) begonnen, die als Vermarktungsgrundlage dienen sollte. Zwölf nicht mehr betriebsnotwendige denkmalgeschützte ehemalige Abspannwerke wurden anhand von historischen und aktuellen Fotos und Bauplänen beschrieben. Denkmalrelevante Erhaltungsbereiche wie die im Verbund gemauerten Fassadenabwicklungen aus Backstein, die Schaltwarten mit ihren Marmortafeln und die Erschließungsbereiche wurden beschrieben und grundsätzliche Modifikationsmöglichkeiten für die geplanten Umnutzungskonzepte dargestellt.

Die vom Eigentümer und von der staatlichen Denkmalpflege mit Engagement betriebenen Umnutzungskonzepte werden durch verschiedene externe Faktoren bestimmt. Die gesamtwirtschaftlichen Rahmenbedingungen in Berlin, Finanzierungskonzepte von potenziellen Käufern, vorhandene Wachstumsperspektiven der möglichen Nutzer und planungs- und baurechtliche Forderungen der Genehmigungsbehörden hatten auf die erarbeiteten Nachnutzungskonzepte Einfluss. Die Konzepte zur Umnutzung von denkmalgeschützten Sondergebäuden bestehen daher im Wesentlichen aus wirtschaftlichen Bewertungsstrategien und planerisch umsetzbaren Nutzungsstrategien.
Ehemalige Kraftwerke bieten Räume für Ausstellungskonzepte, Freizeit- und Gastronomiekonzepte, aber auch für Handel und Produktion. Die städtebauliche Lage und Einbindung spielte für eine zeitnahe Umsetzungsmöglichkeit von Umnutzungskonzepten eine wesentliche Rolle, da der Berliner Immobilienmarkt seit Mitte der 90er Jahre in mehrere Teilmärkte aufgeteilt ist, die sich sehr unterschiedlich entwickeln.
Die Bewag entschied sich frühzeitig dafür, auch temporäre Nutzungen in ihren Gebäuden zu fördern. Dies geschieht vor allem, um die bisher im Wesentlichen unbekannten Standorte einer breiteren Öffentlichkeit

Ever since 1997, the Bewag's aim regarding land transactions has been to apply targeted re-utilisation schemes and marketing measures to create a market for listed buildings of special interest to the electrical industry, and to sell these buildings to owners who make use of them personally, or else rent them out. Together, the Bewag, the Regional Authority for Monument Preservation and freelance architects have jointly worked out a new form of monument-management: working out imaginative ways to maintain these listed buildings of technological importance. The areas that are to be preserved are well documented; and there is ample scope for further developing the buildings and their potential for additional or new tasks.

This constructive process was started with the publication: "Electropolis Berlin" (Published by the Bewag, Berlin 2000), which should serve as the basis of a marketing strategy. The book uses both historical and contemporary photographs as well as building-plans to describe twelve former power installations which are no longer needed for current operations but are preserved monuments. Areas of listed buildings that need to be preserved, such as the brickwork façade-corners in bonded stonework, the switching stations with their marble boards, and potential developments, were all described, as were basic modifications for proposed re-utilisation schemes.

The re-utilisation projects that have won a firm commitment both from the owner and the State Authority for the Preservation of Monuments are determined by various external factors. The overall economic conditions in Berlin, financial projections by potential purchasers, the existing growth prospects of potential users, as well as licensing authorities' regulations for planning and building – all these influenced existing proposals for re-utilisation. In other words, plans to re-use listed special buildings consist largely of economic revaluations and realistic planning strategies for re-utilisation. Former power stations offer locations for exhibition projects, leisure and gastronomy, as well as for commerce and production. The state of current and future municipal building in Berlin was crucial in creating an imminent opportunity to put re-utilisation proposals into practice, as since the mid-90s, the Berlin property market has split into several subsidiary markets that are developing in very different ways. At an early stage, the Bewag decided to encourage even temporary re-utilisation in their buildings. The

Kraftwerk Rummelsburg, Verfallszustand Kesselhaus, Aufnahme 1998
Rummelsburg Power Station, boiler-house in a state of decay, photograph 1998

bekannt zu machen. Da von den Abspannwerken während ihrer aktiven Betriebszeit keine umfeldrelevanten Belastungen ausgegangen waren, hatte man auch deren Außerbetriebnahme nicht bemerkt. Die Ansiedlung des Vitra Design Museums im ehemaligen Abspannwerk Humboldt förderte die spezifische Standortqualität. Fast 100.000 Besucher im Jahr finden ihren Weg zum Vitra Design Museum im Bewag-Abspannwerk im Prenzlauer Berg und fördern so die Bekanntheit des Ortes. Die Bewag strebt an, auf weiteren Liegenschaften Unternehmen und Institutionen anzusiedeln.
Ein weiteres wesentliches Kriterium für die Umnutzung von denkmalgeschützten Industriebauten bilden die wirtschaftlichen Erhaltungsstrategien. Nach Änderung der Steuergesetzgebung mit Anhebung der steuerlichen Absetzbarkeit von Investitionen in Baudenkmale stieg die Nachfrage nach umnutzungsgeeigneten denkmalgeschützten Immobilien. Trotz dieser Investitionsanreize müssen aber weiterhin die Parameter zur langfristigen Wirtschaftlichkeit erreicht werden. Wesentlicher Faktor zur Senkung der Baukosten waren Planungskonzepte, die einen umfangreichen Erhalt der vorhandenen denkmalgeschützten Substanz vorsahen und auf größere konstruktive Eingriffe verzichteten. Diese Planungskonzepte schafften durch gezielte Eingriffe und Ergänzungen langfristig nutzbare Flächen, die eine hohe Nutzerakzeptanz hatten. Die besondere Qualität der Kraftwerke, die in der Materialität des Baustoffes Backstein und im konstruktiven Konzept des Stahlskelettbaus liegt, half mit, diese Ziele zu erreichen.

Einen anderen Weg zum Umgang mit einem unter Denkmalschutz stehenden Technikgebäude stellte das Konzept für das Kraftwerk Rummelsburg dar. Der Gebäudekomplex war durch Kriegsschäden und mangelnde Instandhaltung zu DDR-Zeiten in wesentlichen Bereichen ab Ende der 80er Jahre baufällig. Durch das Erreichen einer Teilabrissgenehmigung für das einsturzgefährdete Kesselhaus bei einer gleichzeitigen Sicherung der erhaltungsfähigen Bausubstanz des ehemaligen Maschinenhauses und des Erweiterungsbaus des Kesselhauses wurde eine mittelfristige Perspektive erreicht, um funktionale und wirtschaftlich vertretbare Nachnutzungskonzepte zu erarbeiten.

main reason behind this was to make a wide cross section of the public, who hardly knew anything about the sites, familiar with them. Because the installations that are now closed had made no additional burdens on the surrounding areas whilst in active service, people had not noticed that they had ceased operations. The decision by the Vitra Design Museum to move into the mothballed Humboldt Power Station in Prenzlauer Berg enhanced the specific quality of the site concerned. Nearly 100,000 visitors a year make their way to the Vitra Design Museum in the former Bewag power station in Prenzlauer Berg, thus making the place better known. The Bewag continues its efforts to persuade enterprises and institutions to move into other idle installations.
A further vital criterion for re-utilising listed, industrial buildings are strategies to ensure economic sustainability. After tax laws were changed to highlight the tax-free status of investments in listed buildings, there was an increased demand for listed property that was ripe for re-utilisation. That said, despite these incentives to invest, the requirements for long-term economic viability still have to be satisfied. Vital factors for decreasing building costs were planning proposals based on a projection that most of the listed properties still intact would be retained and that major reconstruction work would be avoided. By a few well-chosen interventions and additions, these planning proposals made several areas useful in the long-term that were readily accepted by prospective users. The special quality of the power stations, which lay in the condition of the brick as a building material and the design-concept of steel-skeleton construction, helped achieve these objectives.

Another approach to listed buildings with a technological history was that followed by Rummelsburg Power Station. From the late 1980s onward, several important sections of the complex of buildings were close to collapse as a result of war-damage and inadequate maintenance during GDR days. Obtaining a partial demolition order for the unsafe boiler-house, – at the same time as strengthening sections of the former machine shop that were still sound and the boiler-house extension –, offered an interim solution that allowed time to devise ideas for re-utilisation that would meet both functional and financial considerations.

Kraftwerk Steglitz,
Maschinenhalle,
Innenaufnahme um 2000
Steglitz Power Station,
machine hall,
interior view around 2000

Wertsteigerung durch Denkmalmanagement

Handlungsbedarf besteht für ehemalige Kraftwerke, die durch ihre Hallenarchitektur kaum für konventionelle Umnutzungen zu Verwaltungs- oder Wohngebäuden geeignet sind. Eine Umnutzung zu Kultur-, Museums- oder Ausstellungszwecken scheiterte bisher meist an der angespannten Haushaltssituation der öffentlichen Hand. Hier besteht grundsätzlich Handlungsbedarf dahin gehend, dass mit Priorität die Umnutzungen von denkmalgeschützten Gebäuden gefördert werden sollten, statt weitere Neubauten zu errichten. Ein schonungsvoller und umweltgerechter Umgang mit den Ressourcen Boden und Gebäude könnte das Ergebnis sein.

Nachdem es der Bewag durch das entsprechende Immobilien- und Denkmalmanagement gelungen ist, einen Markt für nicht mehr betriebsnotwendige denkmalgeschützte Umspannwerke zu schaffen, steht diese Aufgabe für ehemalige Kraftwerksareale noch an. Durch die Erstellung der entsprechenden Konzepte für die zukünftig nicht mehr betriebsnotwendigen Teile der Kraftwerke Steglitz, Charlottenburg und Moabit sollen diese Voraussetzungen geschaffen werden. Die wirtschaftlichen Ziele des Substanzerhaltes, der Erlös- und Ertragserwirtschaftung und der langfristigen Kostenreduzierung sollten dabei ebenso erreicht werden wie der denkmalgerechte Erhalt der bauhistorisch bedeutsamen Industriedenkmale.

How Monument-Management Can Increase Value

A demand exists for former power stations whose "hangar architecture" is hardly suitable for conventional conversions into either administrative or accommodation buildings. To date, all plans to re-utilise the sites for purposes connected with culture, museums or exhibitions have usually foundered on the tight state of public budgets. The principle at issue here is that re-utilisation of listed buildings should take priority over the construction of further new buildings. One positive result might be a circumspect and environmentally friendly approach to the twin resources of land and buildings.

After the Bewag has – by appropriately managing building-land and the monuments on it – succeeded in creating a market for no longer operational listed transformer plants, the same prospect awaits former power station areas. By implementing appropriate proposals for those parts of the power plants at Steglitz, Charlottenburg and Moabit that no longer need to operate, this ought to be possible. The economic objectives of retaining structure and meeting financial targets, by making a profit and reducing costs, should be achievable at the same time as providing a guarantee that the listed sections of significant industrial monuments will be retained.

EINE ZUKUNFT FÜR DENKMALGESCHÜTZTE KRAFTWERKE
A FUTURE GUARANTEED FOR LISTED POWER STATIONS

Grußwort des Landeskonservators
A Word of Welcome from the Regional Head of Conservation

Jörg Haspel

Die Vermeidung von künftigen Denkmal-Leerständen und die Nachnutzung bereits funktionsloser Denkmalimmobilien zählt zu den großen konservatorischen Herausforderungen unserer Zeit. Weitsichtige Strategien der Denkmalumnutzung sind gerade in Berlin und in einem noch nie da gewesenen Maßstab für Bau- und Technikzeugnisse der historischen Industriekultur gefragt. Viele haben in den letzten Jahren vor dem Hintergrund gewaltiger Modernisierungen und wirtschaftlicher Umbrüche ihre funktionelle und damit auch ihre wirtschaftliche Basis für eine sinnvolle Nutzung und Denkmalerhaltung verloren oder sehen einer unsicheren Zukunft entgegen.

Die Denkmale der Elektroindustrie und der Stromwirtschaft vergegenwärtigen bis heute eindrucksvoll den legendären Ruf Berlins als „Elektropolis" unter den Metropolen des 20. Jahrhunderts. Mit seinem industrietypischen Profil teilt der Denkmalort Berlin freilich auch ein Problem, wie es für viele traditionsreiche Industrie- und Denkmalorte charakteristisch ist: Es werden historische Bedingungen und Funktionen bezeugt, die sich überlebt haben. Diese Zeugnisse werden vielerorts nur überdauern können, wenn es gelingt, den mit der abgebrochenen Nutzungskontinuität entstandenen Freiraum mit vitalen Funktionen neu zu besetzen oder eine nachhaltige Umnutzungsperspektive für die Monumente der Industriekultur aufzuzeigen.

Unter den Traditionsunternehmen, die Berliner Technik- und Architekturgeschichte geschrieben und aus ihrer Vergangenheit ein viel beachtetes Denkmalerbe überliefert haben, das infolge aktueller Rationalisierungsschübe und weltweiter Veränderungen für neue Nutzungen zur Disposition steht, ist die Bewag an vorderster Stelle zu nennen. Das Unternehmen hat vergleichsweise früh die sich abzeichnende Herausforderung angenommen und sich anstelle einer defensiven, eher abwartenden Haltung auf eine offensive Handlungsstrategie des

Avoiding the blight of listed buildings standing empty and re-using listed properties already without a function, these are amongst the greatest tasks to face those concerned with conservation in our day and age. Far-sighted strategies for the re-use of listed land are much in demand not only in Berlin but also – more than ever before – they are a yardstick for assessing construction and technology that is the heritage of our historic industrial culture. Over the past few years, against a backdrop of radical modernisation and economic upheaval, many people have lost their functional and also their economic basis for a sensible pattern both of using and preserving listed buildings; or else, they face an uncertain future.

The listed buildings that form part of the electro-industry and the electrical economy in general are a striking reminder of Berlin's legendary reputation as the "electropolis" among twentieth-century metropolises. With its characteristically industrial profile, Berlin, a city of many listed monuments, also shares a problem that is typical of many places that are rich in tradition and listed property: the evidence remains of historical conditions and functions that have outlived their usefulness. In several places this evidence will only be able to survive if a success is made of injecting the now discontinued functions with a vital new emphasis; or else, if a convincing plan is drawn up to re-use the monuments of industrial culture in a new way.

Among the traditional companies to have charted the course of technological and architectural history in Berlin, rescuing a much-valued legacy of listed buildings from the past, one business that takes pride of place is the Bewag. After undergoing current efficiency drives and worldwide changes, Bewag is now in business for projects of innovative re-use. At a relatively early stage, this enterprise accepted the emerging challenge. Rather than taking a defensive, or "wait-and-see" attitude, Bewag has adopted an offensive strategy both on marketing monuments and on how best to re-use them to the benefit of the listed buildings. One positive result was the

Kraftwerk Charlottenburg, Ansicht von der Spree, Aufnahme um 1990
Charlottenburg Power Station, view from the river Spree, photograph around 1990

Denkmalmarketings und der denkmalgerechten Umnutzung besonnen. Ein Ergebnis war die bereits 2000 erschienene Broschüre „Elektropolis Berlin. Historische Bauten der Stromverteilung".

Nun liegt mit „Kraftwerke in Berlin. Das Erbe der Elektropolis" eine weitere Publikation vor, in der stillzulegende oder bereits aus der Nutzung genommene denkmalgeschützte Kraftwerksanlagen vorgestellt werden. Das Zentrum der Veröffentlichung bildet – analog zum Heft über nicht mehr benötigte Anlagen der Stromverteilung – ein Katalog der durch Leerstand gefährdeten und an neue Nutzungen anzupassenden Kraftwerke der Bewag. Die Publikation informiert in systematischer und knapper Form sowohl die interessierte Fachöffentlichkeit als auch potenzielle Investoren und Nutzer über diesen Baubestand, erläutert dessen architektur- und technikgeschichtliche Bedeutung, klärt Standortbedingungen, liefert Angaben zur planungsrechtlichen Situation, zum baulichen Zustand und zur Flächenbilanz und umreißt vor allem Entwicklungspotenziale sowie denkmalpflegerische Besonderheiten und Spielräume für erhaltende Umnutzungen.

Die Publikation versteht sich als Denkmal- und Immobilieninformation zugleich. Das Ergebnis möchte man als eine Art ökonomischer und konservatorischer Werterhaltungsstrategie begreifen, die im Schnittfeld zwischen Immobilienmanagement und Denkmalmanagement dem drohenden kulturellen Niedergang und materiellen Verschleiß dieser Bau- und Technikdenkmale entgegenwirken will, um bereits im Vorfeld kontraproduktive Weichenstellungen oder unnötige Konflikte zugunsten einer Reaktivierung und Revitalisierung dieses kostbaren Erbes zu vermeiden.

Die Bewag war mit dieser produktiven Herangehensweise bei der Umnutzung denkmalgeschützter Abspannwerke bereits sehr erfolgreich. Ich bin überzeugt, dass die vorliegende Veröffentlichung auch im Hinblick auf stillgelegte Kraftwerke einen wichtigen ersten Schritt zur denkmalerhaltenden Nachnutzung und einen aktiven Beitrag zur nachhaltigen Entwicklung dieses Erbes leisten wird.

brochure, published in 1999, entitled: "Electropolis Berlin: Historic Buildings associated with Electrical Distribution".

With "Power Stations in Berlin. The Electropolis Heritage" another publication is now available, in which listed power plant units that are earmarked for closure or have previously been taken out of service, are introduced to the reader. Similar to the booklet about those sites no longer used for electricity distribution, the centre of this publication consists of a catalogue of the Bewag power stations that are in danger of standing idle and that are ripe for redevelopment. In a systematic and concise manner, this publication informs not only "a trade readership" but also potential investors and users about the state of specific buildings, describing their architectural and technological importance; it clarifies site-specific conditions and provides facts and figures for the legal aspects of the planning situation, as well as for the state of the buildings and their relation to surrounding land. Above all, however, the publication outlines both development potential and monument-specific tasks, as well as opportunities for forms of re-use that will safeguard the listed buildings for posterity.

In other words, the publication fulfils a dual role: it offers information about the listed buildings it mentions and the land on which they are constructed. It is tempting to see the end result as a kind of economic and conservationist strategy for upholding cherished values. From a point somewhere between "property management" and "monument management", it endeavours to avert the threat of these historic and technical buildings falling into cultural decline and material attrition, to avoid counter-productive posturings or unnecessary conflicts, in favour of reactivating and revitalising this invaluable heritage.

The Bewag has already achieved great success with this productive approach to reutilising listed buildings that are no longer in operation. I am convinced that this present publication will mark a significant first step – not least in respect of mothballed power stations – towards re-using land and buildings in a way that preserves the past and contributes to a sustained future development of this heritage.

Kesselhaus Schiffbauerdamm, Aufnahme von 1896
Boiler-house Schiffbauerdamm, photograph from 1896

DIE ANFÄNGE
THE EARLY DAYS

Klaus Bürgel

„Een richtjet Jas wird det im Leben nich“, sagten die Berliner, als die ersten Lichtbogenlampen Plätze, Straßen und sogar schon Gebäude erleuchteten. Doch die Bogenlampen sollten schon bald die seit 1828 in Berlin für Helligkeit sorgende Gasbeleuchtung ablösen. Selbst der Gas-Glühstrumpf von Auer von Welsbach, der das Gaslicht ab 1892 erheblich verbesserte, konnte den Aufstieg der neuen Energieart „Elektrizität“ nicht mehr aufhalten.
Viele Versuche zur Einführung des neuen Lichtes wurden gemacht, seit Sir Humphry Davy um 1808 das elektrische Bogenlicht „entdeckt“ hatte. Die 70er Jahre des 19. Jahrhunderts brachten schließlich zahlreiche elektrische Beleuchtungsanlagen hervor. In Berlin wurden zum Beispiel 1878 das Postgebäude und 1879 der Reichstag, das Preußische Abgeordnetenhaus, die Kaisergalerie, der Pariser Platz und der Anhalter Bahnhof elektrisch erhellt.
Besonders die Straßenbeleuchtungen von Potsdamer und Leipziger Platz, der Leipziger Straße und der Friedrichstraße erregten Aufsehen. Die Anlagen dazu wurden 1882 von Siemens und Halske gebaut und betreut: Vier Otto-Gasmotoren mit je 12 PS trieben vier Dynamos an. Und auch die Edison-Glühlampe wurde erstmals in der Straßenbeleuchtung erprobt.

Die eigentliche Erfolgsgeschichte der Elektrizitätsversorgung begann dann mit der Gründung der AEG 1883 und der Gründung der Bewag 1884 durch Emil Rathenau, der als ersten Technischen Direktor Oskar von Miller gewinnen konnte. Die ersten Blockstationen zur Versorgung eines Häuserblockes entstanden im Jahr 1884. Im August 1885 erfolgte die Betriebseröffnung des ältesten öffentlichen deutschen Kraftwerkes in der Markgrafenstraße 44 (540 kW Leistung mit 100 V Gleichstrom). Im März 1886 ging das zweite Kraftwerk in der Mauerstraße 80 (285 kW Leistung mit 100 V Gleichstrom) in Betrieb. Bald folgte als drittes Kraftwerk die Spandauer Straße 49 (1.680 kW Anfangsleistung mit 2 x 110 V Gleichstrom) und als viertes der Schiffbauerdamm 22 (840 kW mit 2 x 110 V Gleichstrom). Im Jahre 1896 wird in diesem Werk der erste 3.000 V-Drehstrom erzeugt, der im Kraftwerk Markgrafenstraße in

"That'll never be a proper gas-light in a million years", typical Berliners said when the first arc lamps lit up squares, streets and even, at this early stage, buildings. Yet, it was these very arc lamps that would soon replace the gas-lights which had been lighting up Berlin since 1828. Even the reflecting hood, powered by gas, invented by Auer von Wiesbach, which, from 1892 onwards, greatly improved gas-light, could no longer hold up the popularity of the new form of energy known as "electricity". Many attempts to introduce this new form of lighting had been made, ever since 1808, when Sir Humphry Davy had "discovered" the electric arc light. The 1870s eventually produced a large number of electrically illuminated sites. In Berlin, for example, in 1878 the Post Office building, then, in 1879, the Reichstag, the Preußische Abgeordnetenhaus (Prussian Chamber of Deputies), the Kaisergalerie (Imperial Gallery), Pariser Platz, and the Anhalter Bahnhof (Railway Station) were all lit by electricity.
The street-lights that lit up Potsdamer and Leipziger Platz, as well as Leipziger Straße and Friedrichstraße excited particular interest. The machinery to power them was built and maintained by Siemens and Halske: four "Otto gas motors" of 12 horse-power each drove four dynamos. In addition, the Edison filament-lamp was first tried out in a street-light.

The real success-story of electricity supplies then started with the foundation, in 1883, of AEG, and the establishment, in 1884, of the Bewag by Emil Rathenau, whom Oskar von Miller appointed as the first Technical Director. The first "block power stations", which supplied a whole block of flats, were created in 1884. These were followed, in August, 1885, by the opening of the oldest public power station in Germany, at 44 Markgrafenstraße, (with an output of 540 kW, with 100 V d.c., or direct current). In March, 1886, the second power station, at 80 Mauerstraße (producing an output of 285 kW with 100 V d.c.) came on stream. The third power station, at 49 Spandauer Straße (producing an initial output of 1,680 kW with 2 x 110 V d.c.) soon followed, as did the fourth, at 22 Schiffbauerdamm (840 kW with 2 x 110 V d.c.). In 1896, this power station produced the first 3,000 Volts of a.c. (alternating current), which was converted into direct current at the Markgrafenstraße power station. The Oberspree Overland Headquarters

Kraftwerk Mauerstraße, Kolbendampfmaschinen mit Innenpoldynamo, Aufnahme von 1891
Power Station, Mauerstraße, Kolbe-steam machines with inner-pole-dynamo, photograph from 1891

Gleichstrom umgeformt wird. Mit der Drehstrom-Überlandzentrale Oberspree in Oberschöneweide (1 MW Anfangsleistung bei 6 kV) nimmt das erste Kraftwerk außerhalb Berlins den Betrieb auf. 1898 eröffnet dann auf einem Hinterhof das Kraftwerk Luisenstraße 35 (2.000 kW Anfangsleistung). Zugleich wird damit die Periode der „kleinen" Kraftwerke der Bewag abgeschlossen.

Das Jahr 1900 ist von besonderem Interesse, und zwar in doppelter Hinsicht: Zum einen durch das neue Kraftwerk der Gemeinde Charlottenburg, zum anderen durch das zweite Drehstromkraftwerk Moabit in Berlin. Mit diesen beiden Kraftwerken beginnt nämlich die Ära der architektonischen Gestaltung der Gebäude; die Kraftwerke werden zu repräsentativen Bauten.

Hier endet der erste Entwicklungsabschnitt der elektrischen Versorgung dicht bebauter Flächen, der zugleich die Bedeutung Berlins für den Fortschritt der Elektrotechnik deutlich macht. In Deutschland waren im Jahre 1890 ca. 12.000 kW Kraftwerksleistung installiert. Allein auf Berlin entfielen davon mehr als 7.000 kW. Diese Zahlen sind als Teil eines umfassenden Zusammenhanges zu sehen: Die Industrie – speziell die Elektroindustrie – wuchs in einem rasanten Tempo. So hat Berlin um 1900 ca. 50.000 Fernsprechanschlüsse, mehr als ganz Frankreich zusammen. Damals galt Berlin als „elektrische Versuchsanlage". Die hier gesammelten Erfahrungen im Bau von Kraftwerken und Schaltanlagen waren von besonderer Bedeutung, weil sie die ersten überhaupt waren. Ihre Bedeutung für die Elektrizitätsversorgung reichte weit über die Grenzen Deutschlands hinaus.

Die wichtigsten Anlagen und der Stolz der Elektrizitätswerke waren damals die Maschinen- und Kesselhäuser. Doch diese warfen zahlreiche Probleme auf: Die Beschaffung ausreichender Wassermengen zur Kesselspeisung erforderte kostspielige Brunnenbauten. Da Kohlenbunker noch unbekannte Einrichtungen waren, wurde die Kohle mit Pferdefuhrwerken herangeschafft, auf den Hof geschüttet und von dort mit Körben in das Kesselhaus getragen. Dort blieb nur ein schmaler Gang für den Heizer, der die Kohle mit der Schaufel in das Feuerungsloch befördern musste. Zur Vermeidung allzu großer Rauch-

for alternating current at Oberschöneweide (producing an initial 1 MW at a voltage of 6 kV) was the first power station to begin operations outside Berlin. Subsequently, in 1898, the power station at 35 Luisenstraße opened in a rear courtyard (with an initial capacity of 2,000 kW). This marks the end of the period of "small" power stations run by the Bewag.

The year 1900 is of special interest for two reasons: one is the new power station constructed in the parish of Charlottenburg; and the other is the second alternating current power station, constructed at Moabit, Berlin. These two power stations signal the start of the era of architecturally impressive constructions. In other words, power stations become imposing buildings in their own right.

This building marks the end of the first stage of development in supplying electricity to densely populated areas and, at the same time, it makes clear how important Berlin was for advances made in electro-technology. By 1890, Germany's power stations had a capacity of roughly 12,000 kW. Berlin alone enjoyed more than 7,000 kW of this total output. These figures should be seen in the context of a much wider picture: industry – and especially the electro-industry – was growing at a fantastic rate. By 1900, for example, Berlin had some 50,000 telephone connections, – more than the whole of France! In those days, Berlin was regarded as the "electrical experimental laboratory" par excellence. The experiences gathered here in constructing power stations and switching facilities were especially significant because they were the first to be built anywhere. Their importance for electricity supply extended far beyond the frontiers of Germany.

In the early days, the most important installations and the pride of electricity plants were the machine and boiler houses. Even so, they each created many problems: supplying sufficient quantities of water to feed the boilers demanded the construction of expensive wells. Because coal-bunkers were still unheard of, the coal was transported on horse-drawn vehicles, tipped out into the yard, then carried to the boiler-house in baskets. Inside, there was only a narrow passageway for the stoker, who had to feed the coal into the fuel-hole by shovel. To avoid excessive smoke, wherever possible, British semi-anthracite coal was used. All the same, numerous complaints were always being addressed to the city magistrates' office. One letter, for

belästigung wurde nach Möglichkeit englische Magerkohle verfeuert. Trotzdem gingen beim Magistrat der Stadt immer wieder Beschwerden ein. So heißt es in einem Schreiben: „Kaum öffnet man die Fenster, dann fliegt der Rahm [Ruß] dicht auf unsere Fensterbretter. Unsere Höfe, die sonst zu wirtschaftlichen Zwecken als zum Wäschespülen usw. gebraucht werden, sind mit Rahm überschüttet und deshalb unbrauchbar." Der Raummangel der viel zu kleinen Grundstücke in der Innenstadt machte eine zweckmäßige Kraftwerksplanung unmöglich. Deshalb wurden nur stehende Dampfmaschinen verwendet und die Kessel im Allgemeinen über den Maschinenhäusern angeordnet. Lärmschutzmaßnahmen waren noch unbekannt. Die Maschinen standen auf hölzernen Fundamenten und die Riemenscheiben waren bestenfalls mit Filz und Leder belegt. Kein Wunder also, wenn der Beschwerdeführer seine Klage fortsetzt: „Die Erschütterungen, die das Haus zu erleiden hat, sind so gewaltig, dass die Fußböden, Treppen und Türen zittern, das Geschirr klirrt und Lampen und Kronen sich bewegen. Die heißen Dämpfe durchziehen weit das Erdreich. Die Keller sind durch die Wärme und durch die Erderschütterungen untauglich geworden. Was haben wir verschuldet, um von so großem Unglück heimgesucht zu werden?"

Die elektrischen Anlagen des ersten Berliner Kraftwerkes erscheinen heute natürlich unvorstellbar primitiv. Da auf keinerlei Erfahrungen zurückgegriffen werden konnte, mussten die Mess- und Schalteinrichtungen erst erdacht und entworfen werden. Die notwendigen Einrichtungen wurden dann oftmals mehr zusammengebastelt als konstruiert. Wichtig war lediglich, dass sie ihren Zweck erfüllten. Eine Berliner Zeitung schrieb 1885: „Die in den Dynamos erzeugten kontinuierlichen und gleichgerichteten Ströme werden durch starke Kupferstücke in den Schaltraum geführt, der den interessantesten Teil der Anlage ausmacht, der auch eine gewisse Geschicklichkeit und Ruhe der Bedienung voraussetzt." Die Schaltanlage bestand ursprünglich aus einem Eisengerüst mit hölzernen Tafeln, auf der Sammelschienen gehalten und Messgeräte sowie Sicherungen angeschraubt waren. Daneben standen die Maschinenregler zur „bewussten Regulierung", d.h. zur handbetätigten Feldregelung. Für die Parallelschaltung der Generatoren bediente man sich einer Lampenbatterie. Mit ihr wurde die zuzuschaltende Maschine zunächst belastet, bis sie die Spannung der in Betrieb befindlichen Generatoren erreicht hatte. War die Parallelschaltung geglückt, wurde die Lampenbatterie stufenweise wieder abgeschaltet. Die Lampenbatterie war das wichtigste und kostbarste Hilfsgerät. Da 1885 eine Glühlampe von 16 Normalkerzen 6 Mark kostete, stellten alle Lampen der Batterie einen Wert von 3.000 Mark dar. Zum Vergleich: Ein Arbeiter verdiente damals ca. 20 Mark in der Woche.

Aus dieser Anfangszeit, als Berlin aufbrach, zur „Elektropolis" zu werden, sind uns keine bemerkenswerten technischen Einrichtungen erhalten geblieben. Auf dem Grundstück des ersten öffentlichen Kraftwerks in der Markgrafenstraße 44 wurde wieder gebaut: Die Bewag errichtete hier in den 1990er Jahren ein repräsentatives Büro- und Wohnhaus, das „Emil-Rathenau-Haus". Auf den Grundstücken Mauerstraße 78-80 und Schiffbauerdamm 21 sind jedoch noch anschauliche Bauteile als älteste Zeugnisse der Elektrizitätswirtschaft Deutschlands zu finden.

instance, said: "You only have to open the window and the 'cream' (= rust) settles thickly on our window-sills. Our yards, which are used for other, more productive purposes than being repeatedly washed, are coated with 'cream', which makes them unusable." The shortage of space in the far too small inner-city sites made it impossible to plan the power stations properly. Accordingly, only free-standing steam-machines were used and the boilers were generally installed above the machine-houses. Anti-noise measures were still unknown. The machines stood on wooden foundations and the belt-pulleys were at best covered in felt and leather. No wonder, then, that the complainant goes on: "The shaking that the house has to endure is so massive that the floorboards, staircases and doors tremble, the crockery clatters, and lamps and casings shudder. Blasts of hot steam penetrate far and wide. The heat and the shaking of the earth all around have made the cellars unusable. What crime have we committed to be cursed with such dire misfortune?"

Today, naturally enough, the electrical installations of the first Berlin power station look unimaginably primitive. As there was no prior experience to fall back on, the mechanisms for measuring and switching current first had to be invented and designed. As often as not, any necessary equipment was then more cobbled together than constructed. All that really mattered was that it did its job. A Berlin newspaper wrote in 1885: "The direct and alternating electrical currents generated in the dynamos are led via massive pieces of copper into the switchgear room. This is the most interesting part of the whole installation and calls for a certain dexterity and presence of mind in operation." Originally, the switching facilities consisted of an iron frame with wooden panels, onto which current collectors, measuring instruments and fuses were attached. Beside this stood the machine-regulators for "conscious regulation", in other words, for a manually operated rheostat. A lamp-battery was used to connect the generators' circuits. This was also used to give the machine an initial charge, until it had reached the voltage of the operating generators. Once the circuit connection had been successfully established, the lamp-battery would be gradually switched off again. The lamp-battery was the most important and most valuable appliance. In 1885, a filament lamp, equivalent of the power of 16 ordinary candles, cost 6 Marks. All the lamps connected to the battery were worth 3,000 Marks. By comparison, a worker at that time was earning about 20 Marks a week.

No significant technical installations have been preserved from these early days when Berlin was just beginning to emerge as the "electropolis". A new building was constructed on the site of the first public power station, at 44 Markgrafenstraße: in the 1990s, the Bewag constructed an imposing building here containing offices and flats, known as the "Emil-Rathenau-Haus". Now, on the sites of 78-80 Mauerstraße and 22 Schiffbauerdamm, there are still visible vestiges of the buildings that represent the oldest traces of electrical plants in Germany.

Verwaltungsgebäude vor der Zentralstation Markgrafenstraße 44 am Gendarmenmarkt, Aufnahme von 1900
Administrative buildings in front of Central Station Markgrafenstraße 44 at the Gendarmenmarkt, photograph in 1900

Das Emil-Rathenau-Haus, ein modernes Wohn- und Geschäftshaus von Max Dudler am Gendarmenmarkt, heute Markgrafenstraße 35, Aufnahme von 1997
The Emil Rathenau House, a modern accommodation and business facility of Max Dudler at the Gendarmenmarkt, today at Markgrafenstraße 35, photograph in 1997

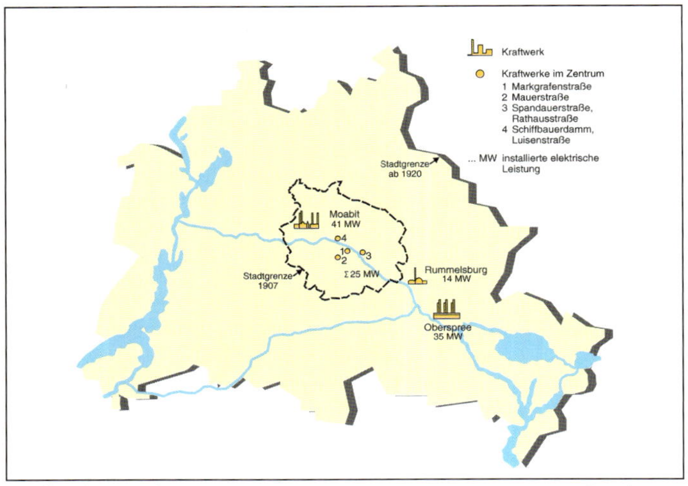

Kraftwerke in Berlin Stand 1907
Power Stations in Berlin Status 1907

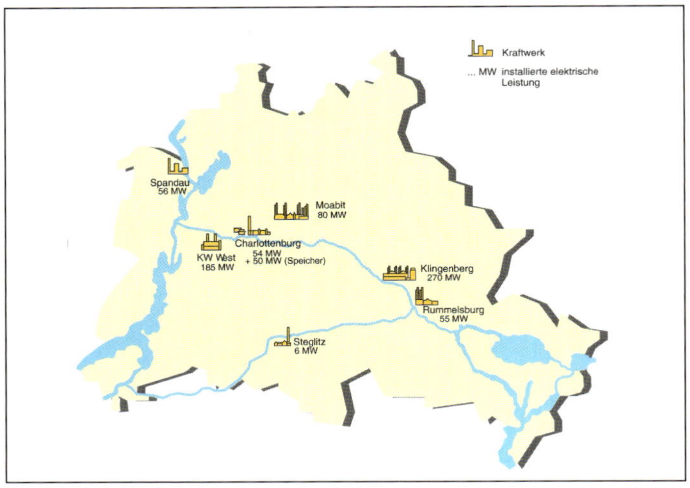

Kraftwerke in Berlin Stand 1931
Power Stations in Berlin Status 1931

AUFBRUCH BERLINS ZUR „ELEKTROPOLIS“
BERLIN'S METAMORPHOSIS INTO AN “ELECTROPOLIS”

Eine Zwischenbilanz im Jahre 2003
An Interim Assessment in 2003

Hilmar Bärthel

Berlin hatte als Großstadt immer auch die Funktion einer Wirtschaftsmetropole. In den letzten zwei Jahrzehnten des 19. und den ersten des 20. Jahrhunderts war Berlin insbesondere geprägt durch die technische Entwicklung und den Aufstieg der Elektroenergie und der Elektrizitätswirtschaft insgesamt. Bekanntlich hatte Thomas Alva Edison 1882 in den USA als technische Neuentwicklung ein erstes privates Kraftwerk zur elektrischen Versorgung der unmittelbaren Umgebung errichtet und strebte weitere in derselben Größenordnung an.
In Auswertung seiner Ideen und Patente verfolgte dagegen der Berliner Politiker und Industrielle Emil Rathenau, unterstützt durch den Ingenieur Werner Siemens, für Berlin von vornherein das Ziel einer öffentlichen und flächendeckenden Versorgung mit der neuen Energie. Die ersten Schritte dienten ausschließlich der Beleuchtung. Später kam der elektrische Betrieb von Straßenbahnen hinzu, fast gleichzeitig wuchsen die Anforderungen der Betriebe für Motoren-Kraftstrom. Im Herbst 1884 wurde in der Markgrafenstraße mit dem Bau des ersten öffentlichen Kraftwerkes in Deutschland begonnen, am 15. August 1885 ging es in Betrieb. Es konnte nur Gleichstrom von 110 V in einen Umkreis von 800 Metern liefern. Der Weg zur Gesamtversorgung der Stadt war weit, der Vision von der „Elektropolis“ Berlin waren durch die verfügbare Technik in diesen frühen Zeitabschnitten enge Grenzen gesetzt. Zur Erzeugung von elektrischem Strom wurde mechanische Energie benötigt. Die konnte anfänglich nur aus Dampfmaschinen gewonnen werden. Diese wiederum benötigten voluminöse kohlengefeuerte Kessel zur Dampferzeugung. Der eigentliche Stromerzeuger, zunächst Dynamo, später Generator genannt, wurde anfänglich als gesonderte Nutzmaschine über Riemen angetrieben, aber schon ab 1890 mit der Dampfmaschine direkt gekoppelt. Ab etwa 1910 kamen zeitweise auch Dampfmaschinen mit beiderseitigen Generatoren zum Einsatz. Die anfänglich eingebauten Generatoren im ersten Kraftwerk in der Markgrafenstraße erbrachten eine Leistung von 45 kW, für die Gesamt-

As well as being a major city, Berlin has always fulfilled the role of an “economic metropolis”. In the last two decades of the nineteenth century and the first two of the twentieth, Berlin was noted for its new technology, especially in the form of electro-energy and the electricity industry as a whole. It is well known that, in 1882, Thomas Alva Edison built the first private power station in the USA. It was a technical innovation that supplied electricity to Edison's immediate neighbourhood and he aimed to build others on the same scale.
Taking full advantage of Edison's ideas and patents, the Berlin-based politician and industrialist Emil Rathenau, supported by the engineer Werner Siemens, sought from the very start, yet in a different way, to make this new form of energy available on a widespread public basis. At first, all their efforts were devoted to forms of lighting. Later on, tramcars were to be powered by electricity, and at about the same time, an increasing number of industrial plants demanded the energy that electric engines could provide. In autumn 1882, in Berlin, work began on Germany's first public power station at Markgrafenstraße and this plant went into operation on 15 August 1885. It could only supply 110 V direct current electricity to a surrounding area of up to 800 metres. Supplying the whole city was a long process. The vision of Berlin as an “electropolis” was strictly limited in the early days because technological advances moved very slowly at first. Generating electrical current required mechanical energy. To start with, this could come only from machines that produced steam. These, in turn, needed large-scale, coal-fired boilers to produce the required quantities of steam. The actual producer of electricity, initially known as a dynamo and subsequently as a generator, was originally a separate servicing machine driven by belts; but as early as 1890, this machine was directly connected to the steam-machine. From about 1910, steam-machines with reciprocal generators were also occasionally deployed. The generators built into the earliest power station at Markgrafenstraße produced an output of 45 kW, – 12 generators were needed for the station's total output of 540 kW. The development of

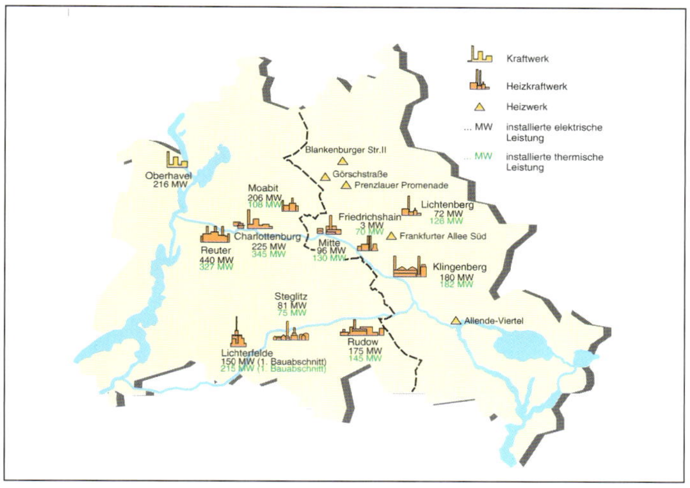

Kraftwerke in Berlin — Stand 1972
Power Stations in Berlin — Status 1972

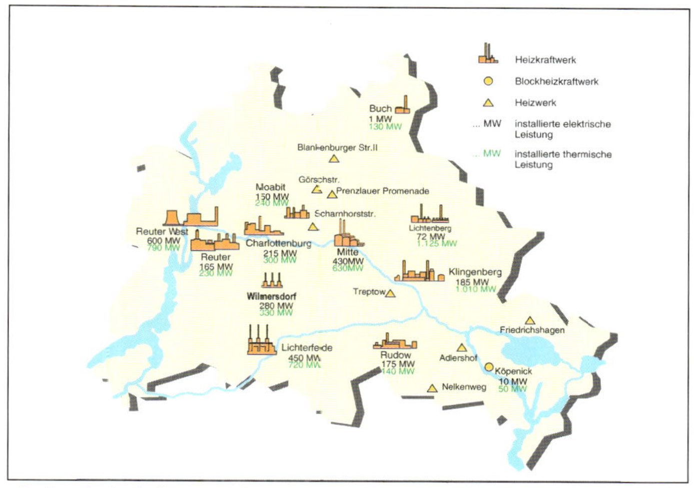

Kraftwerke in Berlin — Stand 2002
Power Stations in Berlin — Status 2002

Kraftwerk Markgrafenstraße um 1885
Markgrafenstraße Power Station, around 1885

Innenansicht des Kraftwerkes Markgrafenstraße, Zeichnung: Kircher 1889
Interior View of Markgrafenstraße Power Station, Drawing: Kircher 1889

Das erste Drehstrom-Kraftwerk Oberspree, Zustand 1905
The first Alternating Current Power Station Oberspree, condition in 1905

leistung des Kraftwerkes von 540 kW waren also 12 Generatoren erforderlich. Die Entwicklung immer leistungsfähigerer Generatoren bedingte die Konstruktion immer stärkerer Dampfmaschinen, für die Dampferzeugung mussten Anzahl, Größe und der Druck der Kessel vergrößert werden. Die bis 1898 in Berlin gebauten sechs kleinen Kraftwerke erzeugten bis 1895 ausschließlich Gleichstrom und mit einer geringfügigen Ausnahme blieb das bis 1926 so. Nur im Kraftwerk Schiffbauerdamm arbeiteten zwei Dampfmaschinen mit vier Generatoren, die Drehstrom herstellen konnten. Die Gleichstrom-Versorgungsbereiche waren wegen der Übertragungsverluste bei längeren Kabeln räumlich begrenzt. Die angestrebte flächendeckende Versorgung erreichte also nur das Kerngebiet der am Ende des 19. Jahrhunderts noch relativ kleinen Großstadt. Neue technische Erfindungen brachten dann die Voraussetzungen zu großen Entwicklungen – „Riesenschritte" bezüglich der Elektrotechnik möchte man sie nennen. Die wichtigsten davon sollen im Folgenden skizziert werden:

Die Jahre von 1884 bis 1899

Es zeigte sich schon bald, dass der Gleichstrom den steigenden Bedürfnissen nicht mehr gewachsen war. Der spätere Chefkonstrukteur der AEG, Michael von Dolivo-Dobrowolski, hatte ab 1894 das Großkraftwerk Oberspree projektiert, das erstmalig den noch heute gebräuchlichen Drehstrom mit einer Spannung von 6.000 V produzierte, 1897 mit einer Anfangsleistung von 4,9 MW in Betrieb ging und 1899 von der Stadt Berlin erworben wurde. Gleichzeitig entwickelte er einen Drehstrommotor und war an der Schaffung weit reichender Kabelnetze für die neue Energieart beteiligt.

Ab 1899/1900

Die Stadt Berlin projektierte und baute mit dem Kraftwerk Moabit ihr zweites Drehstromkraftwerk im Norden der Stadt. Dieses ging schon 1900 mit einer Kapazität von 7,2 MW in Betrieb. Ab etwa 1899 erfolgte zur Senkung des Brennstoffbedarfs der Übergang zum Heißdampfprozess, dessen technische Parameter bis in die Zeit nach dem Ersten Weltkrieg im gesamten deutschen Kraftwerksbau Geltung hatten. Ab 1900 wurden schrittweise Wander- oder Kettenroste zur Erhöhung der Kesselleistung entwickelt und eingeführt. Die Rostflächen konnten erheblich vergrößert werden, die schwere Muskelarbeit zur Kessel-

increasingly efficient generators made it necessary to build more and more powerful steam-machines. The production of steam dictated that the number, size and pressure of the boilers all had to be increased. Until 1895, the six small-scale power stations built in Berlin up to 1898 produced only direct electrical current, and with a slight exception, this remained the case until 1926. Only at the Schiffbauerdamm Power Station were two steam-machines operating four generators that could produce alternating current. Because of transmission losses, any direct current machinery needing long cables was short of space. Thus, it was not until the late 19th century that plans were fulfilled for widespread supply to reach the inner city of what was still a relatively small metropolis. At that point, new technological developments led to enormous strides, – they might be called "giant steps" in electro-technology. We shall outline the most significant of these steps in the following pages:

1884 – 1899

It soon became clear that direct current was unable to meet rising demands. From as early as 1894, Michael von Dolivo-Dobrowolski, later the chief design-engineer of AEG, had drawn up plans for the Oberspree high-capacity power station, which first began producing 6,000 V alternating current, – still in use today. The station began operating in 1897 with an initial output of 4.9 MW and was then acquired by the city of Berlin. Simultaneously, he developed an alternating current motor and he was instrumental in creating far-reaching cable-networks for this new form of energy.

1899/1900 onwards

With the Moabit Power Station, Berlin planned and built its second a.c. (alternating current) power station in the north of the city. This became operational as early as 1900, with a capacity of 7.2 MW. From about 1899 onwards, in order to decrease fuel-requirements, the stations switched to the superheated steam process, whose technical parameters remained valid within all German power stations until some years after World War One. From 1900, travelling or chain grates were gradually developed and introduced, to increase the output of the boilers. The grate surfaces could be greatly enlarged, removing the need for heavy labour, sheer muscle power, to load the boilers. As a consequence, it became necessary to make

Kraftwerk Moabit I, Vollausstattung mit Dampfmaschinen, um 1906
Moabit I Power Station, full equipment with steam machines, around 1906

Kraftwerk Rummelsburg, Erstausstattung mit Dampfturbinen 1907
Rummelsburg Power Station, first equipment with steam turbines, 1907

beschickung entfiel. Nachfolgend ergab sich die Notwendigkeit, mechanische Förderanlagen zum Kohlenantransport bis an die Kessel heran und zum Ascheabtransport zu schaffen. Die Kohlenförderbahnen wurden hoch über die Kessel geführt und entluden automatisch in über den Kesseln angeordnete Bunker.

Die Jahre von 1902 bis 1905

Ab 1902 begannen in den beiden Berliner Kraftwerken die ersten Versuche mit Dampfturbinen, die von der AEG mit den Professoren Riedler und Stumpf entwickelt worden waren. Um 1904/05 standen funktionssichere AEG-Dampfturbinen zur Verfügung. Ihre Leistungsfähigkeit betrug anfänglich mehr als das Doppelte der Leistung der Dampfmaschinen (ca. 5,6 MW je Turbine) und steigerte sich rasch weiter. Der Platzbedarf dagegen verringerte sich auf ein Viertel.

Die Jahre von 1907 bis 1913

Das neu errichtete Kraftwerk Rummelsburg wurde ausschließlich mit Turbinen von insgesamt 13,5 MW ausgestattet. In städtischen Kraftwerken kamen von nun an bei Neubestückungen nur noch Dampfturbinen zum Einsatz. Im Jahr 1913 gab es in den drei Berliner Kraftwerken 16 Turbinen mit Generatoren von insgesamt 110 MW Kapazität.

Das Jahr 1918

Einen wesentlichen technischen Fortschritt stellte der kurzfristige Bau einer 132 Kilometer langen 100-kV-Doppel-Freileitung zum Bezug von Fernstrom aus dem Bitterfelder Braunkohlerevier dar. Die Freileitung erreichte Berlin 1918 am Kraftwerk Rummelsburg, übertrug anfänglich 40.000 kW und erhöhte damit die Versorgungssicherheit für Berlin bedeutend.

Die Jahre von 1920 bis 1923

Mit der Bildung von Groß-Berlin 1920 veränderte sich die Aufgabenstellung zur elektrischen Versorgung prinzipiell: Berlin zählte nun 3,8 Millionen Einwohner und überschritt 1925 sogar die Viermillionengrenze; die Stadtfläche war sprunghaft um das 13-fache auf 878 Quadratkilometer angewachsen. In den zusammengeführten ehemaligen acht Städten, 59 Gemeinden und 27 Gutsbezirken gab es sechs stadt- oder gemeindeeigene Kraftwerke sowie vertragliche Bindungen

transport available, in order to bring coal to the boilers and to dispose of the ashes. The track-ways to supply coal were extended high above the boilers and emptied automatically into bunkers, located above the boilers.

1902 – 1905

From 1902, in both Berlin's power stations, first experiments were made with steam turbines, developed at AEG by Professors Riedler and Stumpf. By about 1904/05, reliable AEG steam turbines had become available. At the beginning, their capacity amounted to double the output of the previous steam-machines (about 5.6 MW per turbine) and it rapidly increased. As another contrast, they also occupied only a quarter of the space.

1907 – 1913

The newly constructed power station Rummelsberg was fully equipped with turbines of 13,5 MW total output. In urban sites, new power stations were only fitted with steam turbines. In 1913, Berlin's three power stations had 16 turbines, whose generators had a total capacity of 110 MW.

1918

Significant technical progress came about with the construction, at short notice, of a 132 kilometre-long 100 kW double overhead transmission line bringing long-distance electricity from the brown coal district of Bitterfeld. This overhead transmission line reached the Rammelsburg Power Station in Berlin by 1913, and with its initial additional output of 40,000 kW of electricity, it significantly increased reliable supplies to Berlin.

1920 – 1923

The formation of Greater Berlin in 1920 fundamentally changed the criteria on which the city, which now had a population of well over a million people, needed to be supplied with electricity. This city of 3.8 million crossed the four million mark by 1925 and the land it occupied had grown by a factor of thirteen to 878 square kilometres. Within the eight former towns, 56 parishes and 27 estates, there were now six city or municipal power stations, as well as contractual links to electrical power from regional suppliers, some of which continued to exist until 1937. Following an initiative of the city authorities, four of the smaller, less profitable plants, those at Neukölln, Pankow,

Kraftwerk Moabit, Zustand um 1925, vorn Kohlenaufbereitung (H.H. Müller)
Moabit Power Station, condition around 1925, at the front, coal processing (H. H. Müller)

über Stromlieferungen von regionalen Versorgern, die teilweise bis um das Jahr 1937 reichten. Nach einer Strategie der städtischen Behörden wurden vier der kleineren unrentablen Werke, und zwar Neukölln, Pankow, Lichtenberg und Köpenick, sofort stillgelegt und die beiden ausbaufähigen Kraftwerke Steglitz und Charlottenburg zum Weiterbetrieb übernommen sowie kurzfristig mit deren großzügiger Rekonstruktion begonnen. Die vertragsmäßig weiterliefernden Regionalgesellschaften konnten über Stromlieferungsverträge an die 1923 gegründete Bewag gebunden werden. In der neuen Großstadt lag die installierte elektrische Kapazität nun bereits bei über 200 MW. Für die Nutzung des Fernstromes in der ganzen Stadt war zunächst eine 100-kV-Leitung vom Umspannwerk Rummelsburg zum 1921 errichteten „Schalthaus" Friedrichsfelde gebaut worden, danach bis 1923 eine 100-kV-Freileitung durch breite Straßen der nördlichen Innenstadt bis zum Kraftwerk Moabit und weiter verlängert zum Kraftwerk Charlottenburg. Die 1925 errichtete 100-kV-Freileitung Zschornewitz-Spandau führte über das Kraftwerk Charlottenburg bis zum Kraftwerk Moabit. Damit konnte Berlin rund 70 MW Leistung zusätzlich beziehen und die Bewag war in der Lage, alle Anforderungen großer Strommengen, beispielsweise für die Elektrifizierung der S-Bahn und für weitere Großabnehmer, zu erfüllen.

Die Jahre 1924 und 1925

Ein technisch bedeutender und technologisch wegweisender Schritt bezüglich der Leistungssteigerung in der Stromerzeugung wurde um 1924/25 im Kraftwerk Moabit mit Versuchen und der nachfolgenden Einführung der Kohlenstaubfeuerung getan. Um die inzwischen betriebenen Hochleistungsturbinen mit ausreichend Dampf zu versorgen, reichten auch die größten Kessel nicht mehr aus. Mehr Dampf konnte aber nur erzeugt werden, wenn ein Mehrfaches an Kohle verbrannt wurde. Das war technisch nur mit Kohlenstaub möglich, der auch Anteile minderwertiger Kle enthalten konnte. Im Kraftwerk Moabit entstand, in Anlehnung an 1924 durchgeführte Versuche, um 1925 nach Entwürfen des Chefarchitekten der Bewag, Hans Müller, eine zentrale Mahl- und Aufbereitungsanlage für Kohlenstaub. Das dreistöckige Gebäude mit hohem Erdgeschoss war vollgestopft mit Trockentrommeln, Kohlenmühlen und Förderanlagen zu den zwölf belieferten Kesseln. Moabit war die „Schule" der Kohlenstaubfeuerung für Berlin und weit über die Stadtgrenzen hinaus.

Lichtenberg, and Köpenick were shut down, and the two expandable power stations at Steglitz and Chalottenburg were not only allowed to continue but they also benefited from extensive reconstruction. The regional companies contracted to continue supplying electricity were 1923 combined into the newly founded Bewag. In the new metropolis, electrical capacity already installed now stood at over 200 MW. To take advantage of long-distance electricity throughout the entire city, initially, a 100 kW transmission line was built from the Rummelsburg Transformer Plant to the "switching station" at Friedrichsfelde, built in 1921. Until 1923, a 100 kW overhead power line ran through wide streets in the northern inner city to the Moabit Power Station and it was further extended to the Charlottenburg Power Station. The 100 kW overhead power line, erected in 1925, which ran from Zschornewitz to Spandau passed through the Charlottenburg Power Station and went on to Moabit. This meant that Berlin could draw on approximately 70 MW of extra capacity. The Bewag was able to meet all demands made, as for instance, for the electrification of the S-bahn, and demands from other major consumers of electrical current.

1924 and 1925

In 1924/25 a technically significant and technologically innovative step was taken by the Moabit Power Station to increase electrical output, by experimenting with, then implementing, the process of coal-dust firing. Even the largest boilers were no longer adequate to provide sufficient steam to the highest-performance turbines that were now installed. Yet more steam could be produced only if a large amount of coal were burnt. Technically, this was only possible by using coal-dust, which might also contain traces of inferior coal. In 1925, at Moabit Power Station, following experiments in 1924 and based on designs drawn up by Hans Müller, chief architect at the Bewag, a central coal grinding and coal-dust preparation site was created. The three-storey building with a high ground floor was crammed full of rotary driers, coal-crushers and dispatch units serving the twelve on-site boilers. Moabit was the "school" of coal-dust firing both for the city of Berlin and far beyond the city limits.

1926 and 1927

The new form of firing made it necessary to dispose of the resultant gas fumes. As early as 1926, in Moabit Power Station the first newly developed Siemens electro-filter towers were tested and installed for this purpose.
Within only 15 months, the Bewag constructed the most productive high-capacity power station in Europe, in which, for the first time, coal-dust was used as the principal form of fuel. This was the Klingenberg Power Station, with a total output of 270 MW. For the first time, turbine units, each with an output of 80 MW, were brought into service. The brainchild was Professor Dr. Georg Klingenberg (who died in 1925). The new station was named after him. The architects Werner Issel and Waltar Klingenberg were responsible for the very spacious and highly effective building, a combination of iron skeleton and clinker construction. From 1927 onwards, this high-capacity power station provided two thirds of the electrical power produced in Berlin, and for years it was a "place of pilgrimage" for power station designers from all over Europe, as well as for the inhabitants of Berlin, school children and other interested parties.

1929

Between 1925 and 1927, Charlottenburg Power Station underwent a complete refit, during which the boiler-house was stripped to its foundations and the machine-house to its outer walls. The result was the first, high-pressure power station in Germany, whose aggregate output could be directed centrally. From 1927, it became possible to double this output. From now on, the plant could feed up to 50 MW of electrical current into the urban

Die Jahre 1926 und 1927

Aus der neuen Feuerungsart ergab sich die Notwendigkeit der Rauchgasreinigung. Schon 1926 wurden im Kraftwerk Moabit die ersten neu entwickelten Siemens-Elektrofiltertürme dafür erprobt und eingesetzt. In nur 15 Monaten Bauzeit errichtete die Bewag das leistungsstärkste Großkraftwerk Europas, in dem erstmals ausschließlich die Kohlenstaubfeuerung großtechnische Anwendung fand: das Kraftwerk Klingenberg mit einer Gesamtleistung von 270 MW. Erstmals fanden Turbinensätze von je 80 MW Anwendung. Geistiger Schöpfer war Professor Dr. Georg Klingenberg (gest. 1925), nach dem das Werk benannt wurde. Für die wirkungsvolle Gestaltung der in Eisenskelettbauweise mit Klinkerverblendung errichteten großräumigen Baukörper trugen die Architekten Werner Issel und Waltar Klingenberg die Verantwortung. Das Großkraftwerk erbrachte ab 1927 nahezu zwei Drittel der in Berlin erzeugten elektrischen Leistung und war jahrelang „Wallfahrtsort" für Kraftwerksbauer aus ganz Europa, aber auch für die Bevölkerung Berlins, Schulklassen und andere interessierte Gruppen.

Das Jahr 1929

Das Kraftwerk Charlottenburg erfuhr 1925–1927 eine Totalerneuerung, bei der das Kesselhaus bis auf die Grundmauern und das Maschinenhaus bis auf die Außenwände abgerissen wurden. Es entstand das erste deutsche Hochdruckkraftwerk, dessen leistungsfähige Aggregate zentral gesteuert werden konnten. Ab 1927 konnte die Leistung verdoppelt werden, es war nunmehr eine Stromeinspeisung von bis zu 50 MW in das städtische Netz möglich. In den Jahren 1928 bis 1929 erfolgte dann als technischer Höhepunkt die weltweit erste Schaffung einer Dampfspeicheranlage nach der Erfindung des schwedischen Ingenieurs Dr. Ruths. In einer Batterie von 16 aufrecht stehenden Stahlkesseln mit einer jeweiligen Höhe von 16,4 Metern konnte Überschussdampf gespeichert und in Spitzenzeiten entnommen werden. Für knapp eine Stunde standen so maximal 50.000 kW zusätzlich zur Verfügung. Für die Anlage wurden zwei Speicherturbinen zusätzlich aufgestellt, dafür musste das Maschinenhaus nochmals verlängert werden. Gekrönt wurde das Ganze 1929 durch die Fertigstellung eines Schornsteines von 125 Metern Höhe mit einem oberen Durchmesser von acht Metern, womit das Kraftwerk Charlottenburg das damals höchste Bauwerk in Europa war.

Die Jahre von 1930 bis 1932

Der in kurzer Zeit weiterhin stark angestiegene Strombedarf Berlins zwang die Bewag, schon ab 1929 ein zweites Großkraftwerk zu planen und 1930-1932 zu bauen, das Kraftwerk West. Es sollte im Westen der Stadt liegen und vornehmlich der Deckung des Spitzenbedarfs dienen. Projektiert und gebaut wurde es durch die Siemens-Schuckert-Werke (SSW). Bis 1930 entstand die erste nutzbare Baustufe, bis 1932 das komplette Werk in Kompaktbauweise. Alle Baukörper waren relativ flach gehalten und technologisch so an- oder ineinander gefügt, dass höchste Rationalität beim Bau und im nachfolgenden Betriebsgeschehen erreicht werden konnte. Als technische Neuerung wurden erstmals in Deutschland auf amerikanischen Entwicklungen beruhende neue Unterschubfeuerungen eingebaut, wodurch die höchsten Kesselleistungen der Bewag überhaupt erzielt werden konnten. Es konnte auch minderwertige Kohle eingesetzt werden und es fiel nahezu keine Flugasche an. Zum Transport der großen Kohlenmengen mussten ein eigener Hafen, ein eigener Kohlenbahnhof mit Großbunker und umfangreiche Außen-Transportbänder angelegt werden. Das Kraftwerk West war mit einer Leistung von 185 MW nunmehr das technisch modernste und technologisch rationellste in Berlin.

Kraftwerk West nach Fertigstellung der 1. Baustufe 1930
West Power Station after completion of 1st. building phase in 1930

grid. The technical highpoint of 1928 and 1929 followed: the Swedish engineer, Dr. Ruth, invented the first system in the world to accumulate steam. Within a collection of 16 steel boilers, all standing upright and each 16.4 metres in height, surplus steam could be accumulated, then released at peak periods. This meant that for roughly an hour a maximum of 50,000 kW of additional power remained available. For this unit, two extra storage turbines had to be installed and the machine-house had to be further extended. In 1929, the whole ensemble was crowned by the completion of a chimney, 125 metres high, with a top diameter of eight metres. Charlottenburg Power Station was the tallest building in Europe at the time.

1930 – 1932

The rapid and sharp increase in Berlin's need for electricity compelled the Bewag to plan a second high-capacity power station as early as 1929. Built in 1930 – 1932, this was known as the West Power Station. It was to be situated in the western area of the city and its prime function was to supply peak demand. It was designed and built by the Siemens-Schuckert-Werke (SSW). By 1930, the first usable stage had been built. By 1932, the whole edifice had been completed by the "condensed construction" method (Kompaktbauweise). All the building components were kept relatively flat and slotted against or inside one another in order to maximise efficiency in terms of construction and ultimate operation. One technical innovation was the decision to use – for the first time in Germany – the "under-thrust" fuelling method developed in America, which enabled the Bewag to achieve its highest ever boiler performance. It also allowed the utilisation of inferior coal and almost no ash at all resulted. Transporting the vast quantities of coal required the building of a special harbour, a special railway station with huge coal-bunkers, and extensive external conveyor-belts. With an output of 185 MW the West Power Station was now the technically most up-to-date and technologically most efficient power station in Berlin.

Kraftwerk West nach sowjetischer Demontage 1945
West Power Station after Soviet dismantling in 1945

Die Kriegsjahre von 1939 bis 1945

Die für Rüstungszwecke stark beanspruchte Stromerzeugung in Berlin wurde 1939 durch sieben größere und zwei kleine Kraftwerke mit einer installierten Gesamtleistung von 781,5 MW sicher beherrscht. Damit war die Bewag zum größten deutschen Stromerzeuger angewachsen. Die Kriegsereignisse verursachten einschneidende Veränderungen an diesen Kapazitäten. Die Kriegsschäden an Kraftwerken betrafen - außer in Moabit und Charlottenburg – im Wesentlichen Gebäude sowie Außenanlagen und hatten relativ wenig Einfluss auf die Stromerzeugung. Die am Kriegsende in der Stadt insgesamt nutzbare Leistung betrug jedoch nur noch 77 Prozent der am Kriegsbeginn insgesamt installierten Kapazitäten. Daran waren aber nicht nur die direkten Kriegsschäden schuld, auch die mangelnde Wartung und der über die Kriegsjahre vorangeschrittene Verschleiß bei Kesseln und Turbinen bedingten Leistungsminderungen. Die weitaus größten Verluste an Strom erzeugenden Kapazitäten erbrachten jedoch die unmittelbar nach Kriegsende vorgenommenen Entnahmen oder Demontagen von Hauptaggregaten durch die sowjetische Besatzungsmacht, die als Teil von Reparationen galten. Bis zum 7. Juli 1945, dem Eintreffen der Alliierten, wurden aus sechs der sieben größeren Kraftwerke Hauptmaschinen, vor allem Turbinen, in der Größenordnung von 351,5 MW demontiert und in Richtung Sowjetunion verladen. Das waren 61 Prozent aller am Kriegsende verfügbaren Stromerzeuger. Ebenfalls 55 Prozent aller Kessel wurden demontiert. Hierunter befanden sich auch die drei modernsten Benson-Kessel, die im Kraftwerk West für eine im Bau befindliche Hochdruck-Vorschaltanlage neu installiert worden waren. Der für den Nachkriegs-Wiederaufbau Berlins verfügbare Rest an Strom erzeugenden Aggregaten betrug also im August 1945 noch 225,4 MW. Das waren 39 Prozent der am Kriegsende noch vorhandenen unzerstörten Substanz.

Die Energiesituation im Westteil Berlins

Die Nachkriegsjahre von 1946 bis 1948

Für die Darstellung der weiteren technischen Entwicklung bezüglich der Energieversorgung Berlins ist an dieser Stelle ein kurzer Exkurs in die Zeitgeschichte erforderlich:

Die Stromversorgung hatte sich in den ersten Nachkriegsjahren wieder

The War Years: 1939 – 1945

The great need for electrical current in Berlin for the armaments' industry was met in 1939 by the installation of seven large and two small power stations with a combined output of 781.5 MW. This made the Bewag the biggest producer of electricity in Germany. The events of the war caused profound changes to these capacities. Except in Moabit and Charlottenburg, the war damage inflicted on power stations was mostly restricted to buildings and external installations, and thus had little effect on the production of electricity as such. Nevertheless, by the end of the war the total electrical output that could still be used in the city had shrunk to 77 per cent of the capacity available before the war. However, it was not simply the result of direct war damage, but inadequate maintenance and the general wear-and-tear that accumulated in boilers and turbines during the war years also took its toll on output. The greatest losses in power resulted directly from the Soviet occupation at the end of the war. The occupying Soviet forces either removed or dismantled major power stations, considering them part of war reparations.

By 7 July 1945, when the Allies arrived, six out of seven of the larger power stations had had major machinery – principally turbines with a total capacity of 351.5 MW – dismantled and carted off to the Soviet Union. This amounted to 61 per cent of all the electricity generated at the end of the war. Similarly, 55 per cent of all the boilers were dismantled. They included the three most up-to-date Benson boilers that had just been installed in the West Power Station for a new high-pressure superposed plant that was about to be built. All in all, in August 1945, the combined electrical capacity available to reconstruct Berlin totalled merely 225.4 MW. This accounted for 39 per cent of the power still available intact at the end of the war.

The Energy Situation in the Western Part of Berlin

The Post-War Years: 1946 – 1948

To explain the further technical development of energy supply in Berlin, it is necessary to make a short excursion into contemporary history. In the immediate postwar years, the provision of electrical power had returned to normal. Supplementary current had been flowing into the western sector

Kraftwerk Charlottenburg nach Umbau 1. Bauabschnitt 1954
Charlottenburg Power Station after reconstruction, 1st. building phase in 1954

normalisiert. In die drei Westsektoren floss Zusatzstrom aus den im Ostteil liegenden Kraftwerken Klingenberg und Rummelsburg sowie Fernstrom aus Zschornewitz über Spandau. Am 19. April 1948 konnte mit dem Wiederaufbau des vollständig demontierten Kraftwerkes West begonnen werden. Nur zwei Monate später, am 26. Juni 1948, verhängte die SMAD, die Sowjetische Militär-Administration Deutschlands, als politisches Druckmittel eine totale Blockade aller Straßen, Wasser- und Schienenwege von und nach den drei Westsektoren und unterbrach alle Stromleitungen. Es entstand eine brisante und wirtschaftlich katastrophale Notsituation, auf die die drei Westmächte mit der kurzfristigen Schaffung einer gigantischen Luftbrücke zur Versorgung von Bevölkerung und Wirtschaft antworteten. Diese und weitere politische Auseinandersetzungen führten im Herbst 1948 zur endgültigen administrativen Teilung der Stadt in einen Westteil, bestehend aus den drei westlichen Besatzungssektoren, und einem aus dem sowjetischen Sektor bestehenden Ostteil, mit jeweils unterschiedlichem Währungsgefüge. Die Spaltung der Bewag in einen West- und Ostteil erfolgte am 6. Dezember 1948. Am 12. Mai 1949 wurde offiziell das Ende der sowjetischen Blockade verkündet. Danach normalisierten sich die Verhältnisse etwas, aber es gab von da an eine politisch und wirtschaftlich absolut isolierte Entwicklung in den beiden Einfluss-Sphären Berlin-West und Berlin-Ost. West-Berlin war durch die Demontagen weit stärker betroffen als Ost-Berlin. Es hatte 77 Prozent aller am Kriegsende noch vorhandenen Kraftwerkskapazitäten verloren, und ein sofortiger Beginn des Wiederaufbaus war auch durch die Alliierten bis 1948 verzögert worden. Verschärfend auf die Situation wirkte sich aus, dass am 5. Mai 1952 die noch bestehenden Leitungsverbindungen zu Ost-Berlin und für den Fernstrombezug endgültig getrennt wurden. West-Berlin war von diesem Tage an zur „Strominsel" geworden. Daraus ergab sich auch die Notwendigkeit, ständig größere Leistungsreserven vorrätig zu halten. Es musste ein gewaltiger und weitsichtiger Prozess zum Wiederaufbau und zur Sicherung der Versorgung von Wirtschaft und Bevölkerung in Gang gesetzt werden. Dieser wurde großzügig durch in- und ausländische Mittel, vor allem durch den Marshallplan unterstützt.

from the power plants of Klingenberg and Rummelsburg in the eastern sector, as well as long-distance power from Zschornewitz via Spandau. On 19 April 1948, a start could be made on reconstructing the completely dismantled West Power Station. Only two months later, 26 June 1948, the Soviet Military Administration of Germany imposed, as a means of applying political pressure, a total blockade of all roads, waterways and railway lines both into and out of the three western sectors, as well as interrupting all the power lines. These steps caused a highly critical and economically disastrous state of emergency, to which the three western powers responded by setting up, temporarily, a massive Luftbrücke ("air bridge") to supply essentials to the inhabitants of West Berlin and its economy. By autumn 1948, this situation and further political confrontations led to the definitive administrative split of the city into a western part, consisting of three western sectors of occupation, and an eastern part, made up of the Soviet sector. Each part had separate currency arrangements. The split of the Bewag into a western and an eastern sector followed on 6 December 1948. On 12 May 1949, the end of the Soviet blockade was officially announced. In the aftermath, the situation normalised a little. However, from then on, political and economic development went their separate ways in the two sectors of West and East Berlin. West Berlin was far more seriously affected by the dismantling of plant than the eastern half of the city. In West Berlin, 77 per cent of all the power-plant capacity existing at the end of the war had been lost and an immediate start to reconstruction by the Allies had been postponed until 1948. The situation was made worse by the fact that on 5 May 1952 the remaining power-lines to East Berlin and those supplying long-distance electrical current were finally cut. From this day on, West Berlin became a "power island". That resulted in a need to have steadily increasing "reserves of output" to hand. What was urgently needed was to institute a far-reaching and far-sighted process that would rebuild and secure power supplies to the economy and the population. This was generously supported by sources both within and outside Germany, notably by the Marshall Plan.

Neubau Kraftwerk Oberhavel um 1963
New Construction of Oberhavel Power Station around 1963

Die Jahre von 1948 bis 1955

Das Hauptereignis dieser Periode war der unter kompliziertesten Umständen erfolgte Wiederaufbau des Kraftwerkes West. Wesentliche Ausrüstungsteile mussten über die Luftbrücke eingeflogen werden. Schon ab Dezember 1949 war eine Leistungsabgabe von 50 MW möglich, im Februar 1950 war die erste Baustufe mit 110 MW betriebsbereit. Ihr folgten 1952 die zweite Baustufe mit 208 MW, 1953 die dritte mit 258 MW und 1954 der vorläufig geplante Endausbau auf 308 MW. Bereits 1953 wurde die neu in Betrieb genommene Spitzenanlage von 50 MW erstmalig mit Fernsehüberwachung für wichtige Betriebszustände ausgerüstet. Um 1955 kamen bei einer Gesamtkapazität von über 500 MW nahezu zwei Drittel des in West-Berlin erzeugten Stromes aus diesem neuen Großkraftwerk, das zu Ehren seines 1953 verstorbenen Förderers den Namen „Kraftwerk Reuter" erhielt.

Im Kraftwerk Moabit wurde erst 1950 die volle Wiederherstellung der kriegsbeschädigten Bausubstanz erreicht. Bis 1953 erfolgte der Abriss der alten Technik bis auf eine Turbine von 24 MW und die erste Baustufe des Neubauprogramms konnte mit 27 MW in Betrieb gehen. Für die Nutzung des niedrigen Kesselhauses mussten lange, liegende Kessel entwickelt werden. Deshalb wurde die zweite Baustufe, bestehend aus einer 55-MW-Heizkraftmaschine und einer 55-MW-Spitzenanlage, erst 1957/58 fertig.

Im Kraftwerk Charlottenburg war die 1929 eingebaute Technik inzwischen weitgehend verschlissen. 1953 wurde daher mit einem Mehrstufenprogramm der Neubau eines modernen Werkes nach dem Prinzip der Kraftwerksblöcke, welches je einen Hochleistungskessel für eine Hochleistungs-Kondensationsturbine vorsieht, begonnen. Die erste Baustufe, bestehend aus einem neuen Kesselhaus mit geringer Grundfläche, 41 Metern Höhe und einer 55-MW-Turbine, war schon 1954, die zweite mit fast gleichen Parametern 1959 betriebsbereit. In diesem Zeitraum begann zudem der Neubau eines großzügigen Fernheiznetzes mit dem Wärmeträger Heißwasser. Außer den Einspeisungen in das Stromnetz konnte auch die Abgabe von Wärmeenergie schrittweise bedeutend erhöht werden.

Das kleine Kraftwerk Steglitz wurde 1958 auf Öl umgestellt und erhielt 1960 als erstes Werk in Berlin eine Gasturbinenanlage mit einer Kapazität von zweimal 25 MW. Die mit vergastem Schweröl betriebenen Turbinen dienten als Spitzenanlage. Sie hatten den Vorteil, nach kurzen Anfahrzeiten hohe Leistungen ins Netz einspeisen zu können.

1948 – 1955

The main event of this period was the complicated process of reconstructing West Power Station. Essential components had to be flown in across the air bridge. Even so, as early as December 1949, it proved possible to supply an output of 50 MW, and in February 1950, the first construction phase was operational, supplying 110 MW of output. This was followed in 1952 by the second construction phase, with an output of 208 MW. The third phase took place in 1953 and produced 258 MW, with the final phase of planned building work in 1954 bringing the total output to 308 MW. As early as 1953, the new peak-period unit, supplying 50 MW, was fitted with CCTV for vital installations. By 1955, out of an overall capacity of over 500 MW, nearly two thirds of electrical current generated in West Berlin came from this new high-capacity plant which, in honour of its designer, who had died in 1953, was given the name: "Reuter Power Station".

In the case of Moabit Power Station, it was not until 1950 that all repairs were completed to the war-damaged building. By 1953, all the old machinery had been dismantled, except for a turbine producing 24 MW. The first phase of the new construction work could start operation, with an output of 27 MW. To make full use of the lower boiler-house long, horizontal boilers had to be developed. This explains why the second construction phase, comprising a 55 MW thermal power machine and a 55 MW peak-period unit, could not be operational until 1957/58.

Over the years, most of the technology built into the Charlottenburg Power Station in 1929 had become obsolete. In 1953, work was started on a modern station, with a multi-step construction programme on the principle of power-plant blocks, each of which includes a high-performance boiler for a high-performance "condensation-turbine". The first phase of construction, comprising a new boiler-house with limited floor-space, a height of 41 metres and a 55 MW turbine, was operational as early as 1954; and the second phase, with almost identical specifications, was ready by 1959. During this same period, work was also begun on a generous long-distance thermal grid, with hot water as the thermal generator. Apart from current fed into the mains, the quantity of thermal energy could be significantly increased step by step.

In 1958, the small Steglitz Power Station was switched to oil and in 1960 it became the first power plant in Berlin to be fitted with a gas turbine, with a capacity of twice 25 MW. The turbines, which were powered by gas-driven heavy oil, served as a peak-period unit. They had the advantage of needing only a short start-up time before feeding high output into the mains.

Neubau Heizkraftwerk Rudow 1963
New Construction of Rudow Thermal Energy Power Station in 1963

Neubau Heizkraftwerk Lichterfelde 1972-1974
New Construction of Lichterfelde Thermal Energy Power Station in 1972 – 1974

Die Jahre von 1961 bis 1963
Auf dem Gelände des alten Kraftwerkes Spandau erfolgte der erste Nachkriegsneubau eines Bewag-Kraftwerkes, des Kraftwerkes Oberhavel. Es war ausgelegt als Grundlastwerk für Stromerzeugung in zwei Baustufen, jede erstmalig mit einem 100-MW-Block ausgerüstet. Eingebaut wurden die im Kraftwerk Reuter bewährten Bensonkessel mit Schmelzfeuerung, die auch mit Gemischen von Stein- und Braunkohlenstaub oder zur Hälfte mit Schweröl betrieben werden konnten. Für alle Aggregate waren Fernbedienung und Fernsehüberwachung vorhanden. Erstmalig erhielt das neue Kraftwerk eine Kompaktwarte, wie sie von da an bei allen Bewag-Kraftwerken Verwendung fand. Das alte Kraftwerk Spandau wurde daneben weiterbetrieben, es erbrachte noch jahrelang seine Standardleistung von 16 MW.

Die Jahre 1963 und 1964
Im alten Teil des Kraftwerkes Steglitz kam zusätzlich eine Heizmaschine mit 25 MW zur Aufstellung. Dazu musste die Leistungsfähigkeit der Kessel bedeutend erhöht werden.

Die Jahre 1964 und 1965
In Rudow ging 1964 der zweite Neubau eines Bewag-Kraftwerkes mit einem 75-MW-Block und 1965 mit einem 100-MW-Block ans Netz. Die Hauptaufgabe dieses Kraftwerkes bestand in der Sicherung der Wärmeversorgung für die Großsiedlung Gropiusstadt. Es konnte aber auch bis zu 75 MW Strom zusätzlich erzeugen. Man musste hier eine Feuerungstechnik ähnlich der im Kraftwerk Oberhavel wählen, da die kostengünstigere Ölfeuerung wegen der Lage im Trinkwasserschutzgebiet nicht angewendet werden konnte. Damit überschritt die installierte Leistung der Bewag erstmals die 1.000 MW-Grenze.

Das Jahr 1966
Im Kraftwerk Charlottenburg wurde die dritte Baustufe des Neuausbauprogramms von 1953 abgeschlossen. Mit dem nunmehr fertig gestellten 75-MW-Kraftwerksblock stieg die Gesamtkapazität auf 225 MW_{el}+345 MW_{th}. Das Kraftwerk erreichte damit den höchsten Wirkungsgrad aller Bewag-Werke bezüglich der gleichzeitigen Abgabe von Strom und Wärme.

1961 – 1963
First new construction after the war was completed on the site of the old Spandau Power Station. This was a Bewag power plant: Oberhavel Power Station. It was designed to supply standard-tariff electricity, in two phases of construction, each initially provided with a 100 MW block. The two Benson boilers retained in the Reuter Power Station were incorporated into the new plant. These were fired not only by melted fuel but could also be fired by a mixture of stone and brown-coal dust or up to half capacity by heavy oil. Remote control and CCTV were made available to all the units. For the first time the new power plant was given a compact-observation tower, as was the standard fitting from now on for all the Bewag power plants. The old Spandau Power Station was still kept in operation as a back-up and for several years it continued to supply its standard output of 16 MW.

1963 and 1964
An additional thermal heater with an output of 25 MW was installed in the old section of Steglitz Power Station. Accordingly, the boilers' potential output had to be vastly increased.

1964 and 1965
In Rudow in 1964, the second new construction of a Bewag Power Plant was connected to the mains with a 75 MW block. The following year, this increased to a 100 MW block. The main task for this power station was to secure the heating supply to the large settlement known as Gropiusstadt. In addition, however, it was able to provide up to 75 MW of extra electrical current. In this case, a fuelling method had to be selected that was similar to the one used at Oberhavel Power Station. The more cost-effective method of oil-firing could not be used in an area where the drinking water supply was protected. With this installation, the total output of the Bewag first crossed the 1,000 MW threshold.

1966
In Charlottenburg Power Station, the third construction phase of the 1953 reconstruction programme was completed. Adding in the now complete 75 MW power plant block, the plant's total capacity increased to 225 MW_{el} + 345 MW_{th}. Thus, the power station achieved the highest degree of effectiveness of all the Bewag power stations supplying electricity and heat simultaneously.

Heizkraftwerk Charlottenburg einschließlich Neubauten für den Umweltschutz um 1998
Charlottenburg Thermal Energy Power Station, including new buildings for environmental protection around 1998

Neubau Gasturbinenkraftwerk Wilmersdorf 1977/1978
New Construction of Wilmersdorf Gas Turbine Power Station in 1977/1978

Das Jahr 1969

Im Kraftwerk Reuter konnte die bisher größte Heizturbine der Bewag mit 132 MW als neue Grundlastanlage in Betrieb genommen werden. Als Dampferzeuger hierfür wurde ein sehr großer Bensonkessel verwendet, der mit 62,5 Metern Höhe 17 Meter über das Gebäudedach hinausragte. Damit wurde nunmehr Reuter zum leistungsfähigsten Bewag-Kraftwerk mit 440 MW.

Die Jahre von 1971 bis 1976

Der Strombedarf von West-Berlin wuchs unaufhörlich weiter. 1971 musste in der Bewag eine Stabsabteilung für die technische Planung der notwendigen weiteren Kapazitäten von Strom- und Wärmeversorgung geschaffen werden. Sie legte fest, in welchen Stadtgebieten weitere Kraftwerke als Grundlastwerke entstehen mussten und wo Anlagen zum Abfangen der Spitzenlasten erforderlich werden, die durch Gasturbinen abgedeckt werden konnten.
Die Bauarbeiten für einen im Süden der Stadt gelegenen vierten Kraftwerksneubau der Bewag in Lichterfelde waren bereits im Gange. Erstmals kamen hier Kraftwerksblöcke von 150 MW zum Einsatz. Durch die Lage in einem Wohngebiet wurden neben Umweltschutzmaßnahmen weitere Einrichtungen erforderlich. So erhielten alle Kraftwerksgebäude eine wirksame Schalldämmung. Als umweltfreundlicher Brennstoff wurde Schweröl eingesetzt, das über Tankschiffe vom Teltowkanal aus störungsfrei angeliefert werden konnte. Für die Rückkühlung des Kühlwassers auf die nunmehr vorgeschriebene Höchsttemperatur von 28 Grad Celsius mussten drei Kühltürme von jeweils 51 Metern Höhe gebaut werden. Die drei Turbogeneratoren gingen planmäßig in den Jahren zwischen 1972 und 1974 in Betrieb. Außer dem Zuwachs von 450 MW_{el} konnten gleichzeitig 651 MW_{th} an Wärmeleistung aus dem Stromerzeugungsprozess ausgekoppelt werden. Bereits ab 1971 wurde durch eine Wärmeleitung zum Kraftwerk Steglitz ein erster Verbund hergestellt, aus dem sich später das Wärmeverbundnetz Süd entwickelte. Die drei Kraftwerksblöcke sind aus Gründen des Umweltschutzes 1988, 1997 und 1998 auf Erdgasfeuerung umgestellt worden.
Im Kraftwerk Moabit erfolgte 1972 zum Abfangen der Spitzenlasten die Inbetriebnahme einer Gasturbinenanlage mit 3 x 17 = 51 MW. Sie wurde in Freiluftbauweise nahe der Altgebäude errichtet.

1969

In Reuter Power Station, the Bewag's largest thermal turbine to date (with an output of 132 MW) was taken into service as the new standard-tariff supplier. Steam was generated by an enormous Benson boiler, which at a height of 62.5 metres tall towered above the building's roof by a full 17 metres. This measure enabled Reuter Power Station to become the most efficient of all the Bewag power stations, with an output of 440 MW.

1971 – 1976

West Berlin's appetite for mains electricity continued apace. In 1971, a special department had to be created at the Bewag to supervise the technical planning of further capacities for the supply of current and heat. It specified in which city districts additional power plants had to be based as generators of standard-tariff supply, and where installations could be sited to take peak demand that could be backed up by gas turbines.
Construction work was already under way on a fourth Bewag power station in Lichterfelde, in the south of the city. Here, for the first time, high-capacity blocks of 150 MW were brought into service. As this power station was sited in a residential area, it proved necessary to institute measures to protect the environment as well as taking other steps. Accordingly, all the power plant buildings were provided with effective sound-proofing. Heavy oil, which could conveniently be shipped by tankers from the Teltow Canal, was chosen as an environmentally friendly form of fuel. In order to cool the cold water down to the newly imposed maximum temperature of 28 degrees centigrade, three cooling towers, each 51 metres high, had to be erected. Between 1972 and 1974, the three turbo-generators came on stream as planned. As well as the increase of 450 MW_{el}, simultaneously 651 MW_{th} of thermal output could also be coupled to the programme of current-production. From as early as 1971, a thermal power line to Steglitz Power Station established the first link that was later to develop into the Wärmeverbundnetz Süd ("Southern Thermal Grid"). To protect the environment, these three power-plant blocks were converted to natural gas fuel in 1988, 1997 and 1998.
In 1972, Moabit Power Station took delivery of a gas-turbine installation, for peak demand, providing 3 x 17 = 51 MW. It was erected as a free-standing, open-air construction unit near the old building.

Heizkraftwerk Reuter einschließlich Neubauten für den Umweltschutz um 1998
Reuter Thermal Energy Power Station, including new buildings for environmental protection around 1998

Die guten Erfolge der Moabiter Anlage bewirkten, dass 1976 im Kraftwerk Charlottenburg eine weitere, deutlich stärkere Gasturbinenanlage mit 3 x 70 = 210 MW angelegt wurde. Diese befand sich aber in einem Gebäude mit Räumen für Warte sowie für technische und soziale Zwecke.
Für die beiden Anlagen Moabit und Charlottenburg musste als Betriebsstoff nunmehr schwefelarmes Leichtöl eingesetzt werden. Für das bisher verwendete Schweröl war aufgrund der hohen Schwefelanteile ein Verwendungsverbot ausgesprochen worden.

Die Jahre 1977 und 1978

Für einen notwendig gewordenen fünften Kraftwerksneubau war schon längere Zeit ein Standort gesucht worden. Die Absicht, mit diesem Kraftwerk zugleich günstige Bedingungen für die Gestaltung eines einheitlichen Wärme-Verbundnetzes zu schaffen, prädestinierte das im Bewag-Besitz befindliche Gelände des ehemaligen Kraftwerkes Wilmersdorf. Da sich auch dort Wohnungen in der näheren Umgebung befanden, wurden – wie schon in Lichterfelde – umfangreiche Maßnahmen zum Schallschutz erforderlich und als Betriebsstoff wurde Öl – nunmehr schwefelarmes Leichtöl – gewählt. So entstand ein höchst modernes, im Städtebauwettbewerb ausgezeichnetes Gasturbinen-Kraftwerk aus drei baugleichen Baukörpern mit je einer Hochleistungsturbine von 92,4 MW, insgesamt also 280 MW Leistungsfähigkeit, sowie einer gleichzeitigen 330 MW Wärmeleistung. Das Kraftwerk ging 1977/78 in Betrieb. Es wurde als Grundlastwerk ausgelegt, kann aber blockweise auch für Spitzenlasten eingesetzt werden.
Damit verfügte die Bewag nunmehr über vier Spitzen-Gasturbinenanlagen und über eine installierte elektrische Kapazität von 2.257 MW für das gesamte West-Berliner Gebiet.

Ab 1986

Die nunmehr folgenden Jahre waren durch die größten finanziellen, technischen und personellen Anforderungen an die Bewag seit den Wiederaufbaumaßnahmen nach dem Krieg gekennzeichnet. Diese hatten ihre ausschließliche Ursache in den verschärften gesetzlichen Bestimmungen für den Umweltschutz. Schon 1983 erarbeitete deshalb die Bewag ein umfassendes Modernisierungsprogramm für die bestehenden Dampfkraftwerke. Man konzentrierte sich dabei zunächst auf die drastische Verminderung der Schwefeldioxid- und Staubemissio-

The great success of the Moabit unit led to the installation, in 1976, in Charlottenburg Power Station, of a further, much more powerful gas-turbine unit, with 3 x 70 = 210 MW. This, however, was sited in a building with observation and other rooms for technical and social purposes.
For the two power plants at Moabit and Charlottenburg the fuel now stipulated was light oil with a low sulphur content. As for the previously utilised heavy oil, because of the very high sulphur content, its use was now banned.

1977 and 1978

A site had long been sought for a fifth new power station. The aim of simultaneously creating favourable conditions for a unified thermal grid made the choice of site a foregone conclusion – on land already owned by the Bewag, the site of the former Wilmersdorf Power Station. Since there were also flats near this site, extensive measures were necessary, and implemented – as in Lichterfelde – for soundproofing; the chosen fuel was oil, now light oil, with low sulphur content. Accordingly, a very up-to-date gas-turbine power plant, awarded a prize in a municipal architecture competition, was developed. It consisted of three similar construction components, each with a high-performance turbine producing 92.4 MW, that is, 280 MW of current, as well as an additional 330 MW of thermal energy. This power plant went into operation in 1977/78. It was built to supply standard power tariffs, but can also be used in a block system for peak demand.
Overall, then, the Bewag now benefited from all four gas-turbine installations for peak demand, and from a combined electrical capacity of 2,257 MW for the whole area of West Berlin.

From 1986

The years immediately following were notable for the greatest financial, technical and personal demands on the Bewag since reconstruction after World War Two. These aspects all had their origin in even stricter legal requirements for the protection of the environment. As early as 1983, the Bewag came up with a comprehensive programme to modernise all remaining steam-driven power stations. Initially, measures concentrated on drastically reducing sulphur dioxide and dust emissions. From 1984, after the conference of Environment Ministers, nitrogen oxide was added to the list

Heizkraftwerk Reuter-West nach Neubau 1987/1989
Reuter-West Thermal Energy Power Station after reconstruction in 1987/1989

Heizkraftwerk Moabit einschließlich Neubauten für den Umweltschutz um 1998
Moabit Thermal Energy Power Station, including new buildings for environmental protection around 1998

nen. Ab 1984 kam nach der Umweltministerkonferenz noch ein Programm zur Reduzierung der Stickoxidemissionen dazu. In den Kraftwerken Oberhavel, Charlottenburg, Reuter und Rudow mussten insgesamt elf Kraftwerkskessel mit Rauchgasentschwefelungsanlagen ausgerüstet werden. Für fünf dieser Blöcke waren zusätzliche neue Elektrofilter erforderlich. Die Realisierung dieser umfassenden Maßnahmen konnte unter erheblichen Anstrengungen bereits in den Jahren 1988 und 1989 abgeschlossen werden.
Für die rund 35 Jahre in Betrieb befindlichen Altanlagen des Kraftwerkes Moabit erwiesen sich zusätzliche Maßnahmen zum Umweltschutz als uneffektiv. Es erfolgte ein Neubau in der Größenordnung von 500 Millionen DM, für den dank internationalen Know-hows das Verfahren der zirkulierenden Wirbelschichtfeuerung (ZWS) angewendet werden konnte. Der Umbau setzte die Tradition des Kraftwerks Moabit und der Bewag als Wegbereiter neuer Technik fort. Verbunden mit einem 100-MW-Kraftwerksblock war das Kraftwerk seinerzeit die größte ZWS-Anlage in Europa. Baubeginn war 1986, die Inbetriebnahme erfolgte Ende 1989.
Nachfolgend wurde bis 1991 in allen übrigen Dampfkraftwerken ein Programm zur Verminderung der Stickoxidemissionen realisiert. Dazu mussten insgesamt 13 Kraftwerksblöcke mit Katalysatoranlagen zur Reduzierung der Stickoxide ausgerüstet werden. Im Ergebnis konnten die Emissionen aller Kraftwerke bei Schwefeldioxid um circa 75 %, bei Stickoxiden um circa 80% und bei Rauchgas um 65 % (= Entstaubungsgrad 99,2 %) vermindert werden.
Die Investitionen für die Gesamtheit dieser Maßnahmen betrugen rund drei Milliarden DM.

Das Jahr 1989

Dem Kraftwerk Reuter wurden bereits seit mehreren Jahren wachsende Mengen an zusätzlichem Dampf aus der Müllverbrennungsanlage Ruhleben zugeführt. 1989 nahm dort der achte Dampferzeuger den Betrieb auf und im Kraftwerk Reuter wurde eine eigene Turbine, die Maschine M, zur energetischen Nutzung des Dampfes installiert. Damit konnte die ehemalige Spitzenanlage stillgelegt werden. Die Kapazität belief sich nunmehr auf 218 MW_{el} und 269 MW_{th}.
Nach jahrelangen Projektierungs- und Bauarbeiten ging mit dem Kraftwerk Reuter-West das nunmehr leistungsstärkste Großkraftwerk der

of banned substances. In Oberhavel, Charlottenburg, Reuter and Rudow Power Stations a total of eleven power-plant boilers had to be fitted with flue-gas sulphur-removal units. For five of these plants, new electro filters also had to be installed. These comprehensive measures were completed – after a considerable amount of work – by 1988/1989.
For Moabit Power Station, which had been in operation for 35 years, other additional steps to protect the environment proved ineffective. Thus, the power station was rebuilt to a design costing DM 500 million. For this reconstruction, international know-how, in the form of circulating layered firing (Zirkulierende Wirbelschichtfeuerung, or ZWS) was applied. The reconstruction work continued the tradition of both Moabit Power Station and the Bewag as pioneers of new techniques. Connected, as it was, to a 100 MW plant, this power station was, at the time, the biggest ZWS installation in Europe. Building work started in 1986 and the plant became operational in late 1989.
Subsequently, until 1991, a programme to limit nitrogen oxide emissions was implemented in all the other steam-driven power stations. To achieve this, a total of 13 power plants had to be fitted with catalytic units to reduce nitrogen oxides. Overall, the measures resulted in the following reductions in all plants: sulphur dioxide was reduced by about 75%, nitrogen dioxide by about 80%, and flue-fumes were down by about 65% (equivalent to a "dust-removal level" of 99.2%).
The total investment for all these measures amounted to around DM 3 billion.

1989

For several years, Reuter Power Station had been fed additional steam from the waste incinerator at Ruhleben. In 1989, the eighth steam-generator began operations there. At Reuter Power Station, an on-site turbine, known as "Machine M", was installed to make energetic use of the steam. This meant that the unit previously supplying peak demand could be shut down. The station's capacity now comprised 218 MW_{el} and 269 MW_{th}.
After years of planning and building work, the Reuter-West Power Station, the sixth new structure in two construction phases, now went into service as the Bewag's most productive high-performance power station. All necessary measures for environmental protection were already in place here. The

Heizkraftwerk Mitte nach Inbetriebnahme 1964
Mitte Thermal Energy Power Station after beginning service in 1964

Heizkraftwerk Lichtenberg nach Inbetriebnahme 1972
Lichtenberg Thermal Energy Power Station after beginning service in 1972

Bewag als sechster Neubau in zwei Bauabschnitten in Betrieb. Hier waren alle Umweltschutzmaßnahmen bereits integriert. Die Nennleistung beträgt 2 x 300 = 600 MW, aus denen maximal noch 2 x 387 = 774 MW an Wärme ausgekoppelt werden können. Der Standort direkt neben dem Heizkraftwerk Reuter ist durch eine hohe Gebäudedichte, riesige Lagerplätze für 355.000 Tonnen Kohlen, über zwei Kilometer lange eingehauste Kohlenförder-Bandstraßen, Silos, Tanks und kilometerlange Rohranlagen gekennzeichnet. Der Neubaukomplex stellt gleichzeitig einen Höhepunkt moderner Industriearchitektur dar, ist mit allen Raffinessen hochwirksamer Umweltschutztechnik ausgestattet und erforderte Investitionen in Höhe von 2,1 Milliarden DM. Soweit absehbar, ist dieses Heizkraftwerk wahrscheinlich der letzte Neubau eines großen Leistungszentrums für Kraft und Wärme in Berlin.

Die Energiesituation im Ostteil Berlins

Die Situation ab 1945

Mit dem zwar veralteten, aber produktionsfähigen Kraftwerk Rummelsburg und dem nur teilweise demontierten Kraftwerk Klingenberg standen im kleineren Teil Berlins rund 65 MW mehr an Kapazitäten zur Verfügung als im westlichen, größeren Teil. Beschränkungen in der Stromerzeugung gab es nur wegen Kohlenknappheit.

Die Jahre von 1945 bis 1951

Im Oktober 1945 wurde bekannt, dass die ausgebauten, noch nicht abtransportierten Teile aus Klingenberg von der Sowjetunion zurückgegeben werden. Deren Wiedereinbau war aber erst 1950 beendet. Das Kraftwerk Rummelsburg konnte bis 1951 auf eine Kapazität von 71 MW ausgebaut werden. Damit traten in der Strombereitstellung einigermaßen normale Verhältnisse ein.

Die Jahre von 1954 bis 1960

Das Kraftwerk Rummelsburg war wegen des Mangels an Steinkohle schon 1954 auf Braunkohle umgestellt und dafür mit neuen Entlade- und Förderanlagen versehen worden. Dadurch gelang es, das Kraftwerk bis 1955 auf eine Leistung von 74 MW zu bringen, die etwa bis 1960 durchgehalten wurde. Danach ging es mit der veralteten Erzeugungs-

nominal output was 2 x 300 = 600 MW, to which a maximum of 2 x 387 = 774 MW of thermal energy could be added. The site directly next to Reuter Thermal Power Plant is notable for a high, wide building, with vast storage spaces for 355,000 tonnes of coal, two kilometre-long, built-in conveyor-belts to move the coal, silos, tanks and miles of pipes. At the same time, this new construction complex also represents a high point of up-to-the-minute industrial architecture. It benefits from all the very latest technology for environmental protection, and required an investment of DM 2.1 billion. For the foreseeable future, this thermal power plant is the last new construction for a large-scale output centre supplying electrical and thermal energy in Berlin.

The Energy Situation in the Eastern Part of Berlin

The Situation from 1945

Allowing for the admittedly outdated, but still productive Rummelsburg Power Station, as well as the partially dismantled Klingenberg Power Station, there were some 65 MW additional capacity available in the smaller half of Berlin than in the larger, western part. Limitations in output were only due to coal shortages.

1945 – 1951

In October 1945, it was announced that the dismantled, though not yet removed components from Klingenberg Power Station were to be given back by the Soviet Union. Even so, they were not finally re-incorporated until 1950. By 1951, Rummelsburg Power Station was expanded to a total capacity of 71 MW. This step marked the resumption of relatively normal power supplies.

1954 – 1960

Due to a lack of bituminous coal, Rummelsburg Power Station was converted to brown coal as early as 1954. For this purpose, it was fitted with new unloading installations and conveyor belts. These measures made it possible to advance the power station to an output of 74 MW by 1955, a level maintained until around 1960. Afterwards, the outmoded production tech-

Heizkraftwerk Lichtenberg,
Zustand 2002
Lichtenberg Thermal Energy
Power Station, condition in 2002

technik rasant bergab und die notwendige Stilllegung war nur noch eine Frage des Termins für einen Ersatzneubau.

Die Jahre von 1960 bis 1964

Besonders kritisch wurde es, als mit dem Neuaufbau des Stadtzentrums ab 1962 der Energiebedarf erheblich anwuchs. Vor allem wurde nicht nur mehr Strom, sondern vor allem auch Wärme für die moderne Beheizung der neuen Wohnungs- und Repräsentativbauten gebraucht. Deshalb musste der Standort eines neuen Kraftwerkes möglichst nahe am Stadtzentrum und natürlich günstig für die Brennstoffanlieferung liegen. Die gesetzlichen Vorschriften zur Luftreinhaltung schlossen den Einsatz von Kohle als Brennstoff aus. Es entstand ab 1960 in vier Jahren das Heizkraftwerk Mitte für den Betrieb mit schwerem Heizöl (Masut). Der Standort am linken Spreeufer nahe der Jannowitzbrücke ließ die Errichtung größerer Ölbehälter nicht zu, deshalb wurden kleinere Behälter für den Vorrat von etwa einer Woche gebaut. Für die notwendige Bevorratung wurde ein Tanklager für 30.000 Kubikmeter direkt an der Spree im 6,5 Kilometer entfernten Kraftwerk Klingenberg errichtet, zu dem laufend drei Ölprahme pendelten.
Die maschinelle Ausstattung von 3 x 32 = 96 MW konnte bei voller Leistung noch 130 MW an Wärme erzeugen, bei Reduzierung der elektrischen Leistung auf 46 MW maximal 284 MW an Wärme. Das neue Heizkraftwerk ging Ende 1964 in Betrieb. Im April 1966 erfolgte dann die Außerbetriebnahme des Kraftwerkes Rummelsburg.

Ab 1970

Der nächste Kraftwerksneubau stand an, als etwa ab 1970 eine Reihe größerer Industriebetriebe aus dem Stadtzentrum in das neu geschaffene Industriegebiet Lichtenberg-Nordost verlagert werden mussten, die Strom und Wärme brauchten. Mit überwiegender Orientierung auf Stromerzeugung wurde das mit Öl betriebene Heizkraftwerk Lichtenberg mit 2 x 36 = 72 MW projektiert und gebaut, das bei voller Stromerzeugung noch 126 MW_{th}, bei maximaler Wärmeauskopplung von 207 MW_{th} jedoch nur noch 24 MW_{el} erbringen konnte. Im Juni 1972 war Inbetriebnahme.
Das Heizkraftwerk versorgte ab Betriebsbeginn die Neubaugebiete im Stadtbezirk Lichtenberg, so auch den Wohnkomplex Fennpfuhl. Es erhielt eine erhöhte Bedeutung, als begonnen wurde, die großen Neubau-Wohnsiedlungen zunächst im Raum Biesdorf-Marzahn zu

nology went rapidly downhill, and the necessary closure was merely dependent upon how long a replacement would take to build.

1960 – 1964

An especially critical stage was reached when, from 1962 onwards, the rebuilding of the city centre (East Berlin) created a huge increase in energy needs. First and foremost, it was not only more electricity but also more thermal energy that was needed for the up-to-the-minute heating systems installed in the new flats and official buildings. For this reason, it was vital for a new power station to be located as close as possible to the city centre and naturally in a favourable position for fuel deliveries. The legal regulations about clean air made it impossible to use coal as fuel. From 1960, within four years, the Mitte (district of Berlin) Thermal Power Plant had come on stream, powered by heavy oil (masout, a form of petroleum). The site on the left bank of the river Spree near the Jannowitzbrücke ruled out building large oil-containers. Therefore, smaller containers were constructed, to hold about a week's supply. To store necessary reserves, a storage tank containing 30 000 cubic metres was built directly on the Spree at Klingenberg Power Station, which was only 6.5 kilometres away, and to and from which oil-lighters regularly sailed.
The machinery installed at this power plant produced 3 x 32 = 96 MW which, at full capacity, could also produce 130 MW of additional thermal energy. If the electrical output were reduced to 46 MW, it could generate a maximum of 284 MW of thermal energy. The new thermal power plant went into operation in late 1964. Then, in April 1966, Rummelsburg Power Station was mothballed.

From 1970

The next new power station was due in 1970, when a series of large industrial concerns had to be relocated from the city centre to the newly created "business park" of Lichtenberg-Nordost (Northeastern Lichtenberg), which needed electricity and heat. With an overwhelming bias toward the production of electrical current, the oil-fired Lichtenberg Thermal Power Plant was planned and built, producing 2 x 36 = 72 MW. At full electrical power, this plant could produce an additional 126 MW_{th}, though fitted for maximum heat-output of 207 MW_{el}, it could produce only 24 MW_{el}. The plant came into service in June 1972.
From the very start of operations, this power plant supplied the newly constructed areas of the Lichtenberg district, as well as the residential complex

Umbau Kraftwerk Klingenberg zum Heizkraftwerk, 1. Baustufe 1974-1981
Reconstruction of Klingenberg Power Station into a Thermal Energy power plant, 1st. building phase 1974 – 1981

Heizkraftwerk Klingenberg, Zustand um 2002
Klingenberg Thermal Energy Power Station, condition around 2002

errichten. 1975 konnte das Kraftwerk durch Zuschaltung der nebenan in Betrieb gegangenen ersten Baustufe einer Müllverbrennungsanlage in seiner Leistung um bis zu 50 MW verstärkt werden. Danach entstand eine kritische Lage, denn die obere Grenze der installierten Kapazitäten war erreicht. In großer Eile wurden als zweite Baustufe drei ölgefeuerte Heißwassererzeuger mit je 116 MW in Freiluftbauweise errichtet, die bereits 1978 betriebsbereit waren. Die darauf folgende Ölkrise löste den Schnellbau langer Erdgasleitungen aus. Im Jahr 1980 erreichte eine neu gebaute Erdgasleitung das Heizkraftwerk und es erfolgte sofort die Umstellung aller Dampferzeuger auf Volllastbetrieb mit Erdgas.
Schließlich musste ab 1982 in dritter und vierter Baustufe ein gesonderter Anlagenblock von nochmals sechs Heißwassererzeugern zu je 116 MW errichtet werden, von denen der letzte erst 1989 in Betrieb ging. Damit hatte das Heizkraftwerk eine Gesamtkapazität von 72 MW_{el} und 1.177 MW_{th} erreicht, die sich bis 2003 nicht verändert hat.

Die Jahre von 1974 bis 1990

Das 1927 in Betrieb gegangene Großkraftwerk Klingenberg hatte nach dem Wiedereinbau der zeitweise demontierten Teile etwa ab 1951 wieder seine volle Kapazität erreicht. Ab 1960 war es unvermeidbar geworden, etwas gegen die großen Mengen an Flugasche zu tun, die die Umwelt in weitem Umkreis verschmutzten: Alle 16 Kessel wurden bis 1964 mit modernen Elektro-Doppelfiltern ausgestattet. Für die Abführung der Rauchgase in größere Höhen entstanden zwei 140 Meter hohe Stahlbetonschornsteine, die acht Stahlschornsteine von 1927 konnten abgerissen werden. Ab 1974 stand dann die Frage einer grundsätzlichen Rekonstruktion des Kraftwerkes vor allem deshalb an, weil die Brennstoffbelieferung mit Steinkohle nicht mehr gesichert werden konnte. Es entstand ein Totalumbau in zwei Baustufen, bei dem nur die nicht direkt mit der Produktion verbundenen Gebäude erhalten blieben.
Die erste Baustufe bis 1981 umfasste den Neubau des Kesselhauses A mit vier Dampferzeugern, die für schweres Heizöl oder Erdgasbetrieb (alternative Fahrweise) ausgelegt waren, sowie das ebenfalls neu gebaute Maschinenhaus mit drei Turbinen zu je 60 MW, also einer Gesamtleistung von 180 MW.
Am 31. März 1981 wurde die Weiterführung der Erdgas-Haupttrasse vom Kraftwerk Lichtenberg aus bis nach Klingenberg fertig gestellt und

of Fennpfuhl. It became even more important once a start was made on constructing large-scale residential districts, initially in the Biesdorf-Marzahn district. The power station was able to increase output up to 50 MW, thanks to the nearby waste incinerator, whose first phase of construction went on stream in 1975. After that, a critical stage was reached, as the upper limit of installed capacities had been reached. In record time, the second reconstruction phase of three oil-fired hot-water installations, each producing 116 MW, was completed in open-air construction, and became operational as early as 1978. The oil crisis that followed prompted the construction of long natural-gas pipelines. In 1980 a newly built natural gas pipeline reached the thermal power plant and immediately all units producing steam were converted to maximum power on natural gas.
Finally, from 1982, during the third and fourth construction phases, a separate plant had to be built, with six more units each producing 116 MW of hot water, though the last did not come on stream until 1989. All told, this power station had reached a total capacity of 72 MW_{el} and 1.177 MW_{th} – totals which have not changed up until 2003.

1974 – 1990

After the reconstruction of its temporarily dismantled components, Klingenberg High-capacity Power Station, which came into operation in 1927, resumed full capacity in 1951. From 1960, it had become essential to do something about the vast quantities of flue-ash that were polluting the environment for miles around. By 1964, all 16 boilers had been fitted with the latest electro-double filters. In order to remove flue-fumes at greater heights two 140 metre-high reinforced concrete towers were erected. This meant that the eight reinforced concrete chimneys from 1927 could be demolished. From 1974, the question of radically reconstructing the power station arose, primarily because it was no longer possible to ensure future delivery of bituminous coal. A complete reconstruction followed in two phases, during which only those buildings not directly concerned with production were allowed to stand.
The first construction phase, lasting up to 1981, was concerned with reconstructing Boilerhouse A with four steam-generators, designed to be fired either by heavy heating oil or natural gas (alternative mode). The machine-house was also newly reconstructed with three turbines producing 60 MW each, a total output of 180 MW.
On 31 March 1981, the extension of the natural gas pipeline from Lichtenberg Power Station all the way to Klingenberg was completed; and as early

Das neue Heizkraftwerk Mitte, Zustand um 2002
The new Mitte Thermal Energy Power Station, condition around 2002

bereits ab dem 1. April erfolgte die Inbetriebnahme der Neuanlagen mit Erdgasbefeuerung.
Nun stand noch die Aufgabe an, die Erdgas-Haupttrasse bis zum Heizkraftwerk Mitte weiterzuführen, um dieses ölgefeuerte Heizkraftwerk ebenfalls auf Erdgas umstellen zu können. Eine elf Kilometer lange Trasse über Landstrecken war schon projektiert und genehmigt. Da beide Kraftwerke an der Spree lagen und die Entfernung auf dem Wasserweg nur 6,5 Kilometer betrug, wurde erstmals eine Erdgas-Hauptleitung anderthalb Meter unter der Sohle eines Wasserweges projektiert und 1982 in der Rekordzeit von nur knapp vier Monaten fertig gestellt.
Im Kraftwerk Klingenberg beendete die zweite Baustufe, die von 1981 bis 1987 dauerte, den Umbau zum reinen Heizkraftwerk. Auf dem Platz des alten Kesselhauses B entstanden als technische Neuentwicklungen zwei 57,5 Meter hohe Hochdruck-Dampferzeuger für den Betrieb mit Rohbraunkohle sowie ein Kohlenbunker. So konnten nunmehr alle Leistungsanforderungen erfüllt werden. Der mengenmäßig große Braunkohlenantransport von bis zu 4.000 Tonnen pro Tag erfolgt auf dem Schienenweg bis Königs Wusterhausen, von dort auf dem Wasserweg bis zum Stichkanal im Kraftwerk.
Seit Aufnahme des Dauerbetriebs im November 1987 arbeiten die Anlagen so zuverlässig, dass die Öl- und Erdgasanlagen im Kesselhaus A in die Notfall-Reserve gestellt und konserviert werden konnten.
Je nach Einsatzorientierung beträgt bei einer Stromerzeugung von 180 MW_{el} die Wärmeabgabe 912 MW_{th}, bei maximaler Wärmeabgabe von 1.008 MW_{th} sind noch 100 MW_{el} möglich.
In den Jahren von 1987 bis 1990 erhielt das Heizkraftwerk zudem eine neue Rauchgasentschwefelungsanlage, durch deren Einsatz alle Grenzwerte für den Umweltschutz eingehalten werden können. Das Heizkraftwerk Klingenberg ist das einzige Kraftwerk im Raum Berlin, das noch heute voll mit Rohbraunkohle betrieben wird und wirtschaftlich vertretbar Strom und Wärme erzeugt.
Die 1989 in den vier Ost-Berliner Kraftwerken insgesamt installierte elektrische Leistung lag bei 350 MW, die Wärmeabgabeleistung bei 2.346 MW.

as 1 April the new units came into operation, fired by natural gas.
The time was now ripe to extend the natural gas pipeline to the Mitte Thermal Power Plant, in order to have this oil-fired thermal power plant also switched to natural gas. An eleven-kilometre long overland location line was already in the planning stage and given the go-ahead. Since both power stations were on the banks of the Spree and were only 6.5 kilometres apart by water, the first plans were laid for a natural gas pipeline 1.5 metres beneath the bottom of the waterway; and in 1982, this project was completed in the record time of only about four months.
In Klingenberg Power Station, the second phase of construction, lasting from 1981 to 1987, completed the plant's reconstruction into a thermal power plant alone. On the site of the old Boilerhouse B, two new technological developments were created, a pair of high-pressure steam generators, 57.5 metres in height, for the unit fired by "brown coal", together with a coal-bunker. This meant that all output targets could now be met. There then followed a large transport of up to 4,000 tonnes per day along the railway line as far as Königs Wusterhausen, and from here, by waterway to the branch canal at the power plant.
Since the beginning of the full-time generators in 1987, the installations had been working so reliably that the oil and natural gas units in Boilerhouse A could be added to the emergency reserves and thus conserved.
According to how the figures are calculated, at 180 MW_{el}, the thermal output generated amounts to up to 912 MW_{th}; and at the maximum thermal generation of 1,008 MW_{th}, it was still possible to generate additionally 100 MW_{el} of current.
Between 1987 and 1990, the thermal power plant was also given a unit to remove the sulphur from flue fumes. The installation meant all the limits applicable to environmental protection could be kept. Klingenberg Power Station is the only power plant in the area of Berlin that is still powered by untreated "brown coal" and produces electricity and heat on an economically viable scale. In 1989, the total output for all four East-Berlin power stations was 350 MW electricity and 2,346 MW thermal energy.

Leitwarte des neuen Heizkraftwerkes Mitte
Leading Observation Tower of the new Mitte Thermal Energy Power Station

Heizkraftwerk Klingenberg, Neue Rauchgas-Entschwefelungsanlage (Flüssigverfahren) ab 1990
Klingenberg Thermal Energy Power Station, new flue-fume and sulphur removal plant (liquid process) from 1990

Die wiedervereinigte Stadt ab 1990

1990 erfolgte die politische Wiedervereinigung Deutschlands und der Hauptstadt Berlin. Im Zuge dieser Ereignisse wurden in kurzer Zeit alle ehemaligen Betriebsteile der Bewag in West und Ost zu einem einheitlichen großen und modernen Energieunternehmen für ganz Berlin zusammengeführt. Der auch nach außen hin deutlich im Stadtzentrum erkennbare Neubau des modernen Heizkraftwerkes Mitte ging 1997 in Betrieb.

Schon unmittelbar nach der Übernahme durch die Bewag war beschlossen worden, das seit 1964 bestehende und nunmehr stark veraltete Heizkraftwerk Mitte stillzulegen und an diesem Standort eine hochmoderne Anlage neu zu schaffen. Dies wurde von September 1994 bis Dezember 1996 realisiert. Es entstand eines der modernsten und effizientesten Heizkraftwerke der Welt – und das an einem Standort, der für die bereits im Bau befindlichen Neubauten am Potsdamer Platz sowie für die Regierungs- und weitere Bauten als ideal gelten konnte.

Für die technische Ausstattung wurde das so genannte „GuD-Konzept" angewendet, eine Hochtechnologie, bei der in einer gemeinsamen Anlage Gasturbinen, Abhitzekessel und eine nachgeschaltete Dampfturbine so miteinander kombiniert werden, dass daraus gleichzeitig maximaler Gewinn an Strom und Wärme resultiert. Der Ausnutzungsgrad des eingesetzten Brennstoffes – in der Regel Erdgas, in notwendigen Fällen leichtes, schwefelarmes Erdöl – beträgt 90 Prozent. Zwei Gasturbinen mit je 185 MW und eine nachgeschaltete Dampfturbine von 108 MW stehen für die Grundlast zur Verfügung. Bei Spitzenanforderungen können noch zwei Heißwassererzeuger zu je 120 MW zugeschaltet werden. Das Erdgas liefert eine auf dem Grund der Spree verlegte Pipeline, für das Heizöl wurde ein Behälter mit 9.000 Kubikmetern Inhalt gebaut. Insgesamt ist die Beheizung und Warmwasserversorgung für 181.000 Wohnungen und 500 öffentliche Gebäude möglich, doch sind diese Größenordnungen gegenwärtig noch längst nicht erreicht. Die Gesamtanlage setzt insbesondere auch durch ihre eindrucksvolle architektonische Gestaltung und aufgrund der technisch geringstmöglichen Umweltbelastungen hinsichtlich Schall- und Rauchgasemissionen internationale Maßstäbe. Die dafür aufgewendeten Investitionen betrugen über 600 Millionen DM. Der Ruf Berlins als „Elektropolis" konnte mit diesem Höhepunkt der Kraftwerkstechnik wieder einmal bestätigt werden.

The Unified City since 1990

1990 saw the political reunification of Germany with the capital city of Berlin. In the wake of these events, within a short space of time, all the former business components in both the western and eastern parts of the Bewag were conflated into one single huge, up-to-the-minute energy conglomerate for the whole of Berlin. The newly constructed Mitte Thermal Power Plant, highly visible in the city centre, came on stream in 1997.

Shortly after the takeover by the Bewag, it was decided to shut down Mitte Thermal Power Plant, which had been in existence since 1964 and was by now seriously outdated and to construct on this site a new, entirely up-to-date installation. This project was completed between September 1994 and December 1996. The result was one of the most modern and efficient thermal power plants in the world; and it was located on what might be called an ideal site for both the new buildings on Potsdamer Platz and the government and other buildings nearby.

To ensure that the power plant had the latest technology, the so-called "G&S" concept was applied. This is a highly developed form of technology. Within the same unit gas-turbines (G), cooling boilers, and an additional steam-turbine (S) are combined together in such a way that the unit produces – simultaneously – maximum output of both electrical current and thermal energy for heating. The overall efficiency ratio of the fuel used, normally natural gas, but when necessary, light oil that is low in sulphur, reaches 90 per cent. Two gas-turbines producing 185 MW each and a back-up steam-driven turbine of 108 MW are available for standard tariff output. At times of peak demand, two further thermal generators, each producing 120 MW, can be switched on. The natural gas is supplied by a pipeline, laid along the bottom of the Spree. A container holding 9,000 cubic metres was constructed for the heating oil. In total, it is possible to provide heating and hot water for 181,000 flats and 500 public buildings, although these maximum figures are far from being reached at present. Due to its impressive architecture and lowest possible levels of pollution to the environment from noise and fume emissions, the whole installation has set new international standards. The total investment made in it has now exceeded DM 600 million. This masterpiece of power-plant technology has re-established Berlin's reputation as the "electropolis".

Die gegenwärtig leistungsstärkste Turbine der Bewag, installiert in 2-facher Ausführung im Heizkraftwerk Reuter-West mit jeweils 300 MW_{el} und 387 MW_{th}
The Bewag's highest capacity turbine to date, installed with twofold capacity in Reuter-West Thermal Energy Power Station, each with 300 MW_{el} and 387 MW_{th}

„Elektropolis" Berlin - eine Bilanz über 119 Jahre

Seit 1884, dem Gründungsjahr des elektrischen Zeitalters in Deutschland und Europa, sind 119 Jahre vergangen. Der Ort war Berlin, hier wurden die ersten Kraftwerke für die Versorgung von Bevölkerung und Wirtschaft gebaut. Die Berliner Elektroindustrie lieferte die ersten Ausrüstungen dafür. Im innovativen Wechselspiel von Technik und angewandter Wissenschaft entwickelte sich die neue Kraft in Riesenschritten und in kurzer Zeit zu einem Leistungsspektrum, das wir heute als selbstverständlich täglich in Anspruch nehmen.
Seit 1884 hat die Bewag einschließlich ihrer Vorgänger insgesamt 27 Kraftwerke gebaut und betrieben. Es entstand also im Durchschnitt alle vier Jahre ein neues Werk. Fünf dieser Werke sind außerdem teilweise mehrfach grundsätzlich rekonstruiert worden.
Anfang 2003 betreibt die Bewag elf hochmoderne Heizkraftwerke und zwei Blockheizkraftwerke mit einer installierten elektrischen Bruttoleistung von rund 2.750 MW sowie eine Vielzahl kleinerer Heizwerke. In allen Heizkraftwerken wird das Brennstoff sparende Verfahren der Kraft-Wärme-Kopplung angewendet, woraus eine Gesamtkapazität an Wärme von etwa 6.000 MW resultiert. Damit wird das europaweit größte Fernwärmenetz mit rund 1.250 Kilometern Trassenlänge und über 13.400 Verteilerstationen gespeist, das rund 600.000 Wohnungen und zahlreiche öffentliche Gebäude und gewerbliche Flächen mit Wärme und Warmwasser versorgt.
Zu jedem Zeitpunkt galten für die Bewag die Anwendung zukunftsweisender Technologien, Initiative und Engagement bei der technischen Entwicklung der Werke und ihrer Ausrüstungen sowie die rationellste Versorgung der Metropole Berlin mit Energie und Wärme als oberste Leitlinien. Von diesen in der Geschichte der „Elektropolis" Berlin fest verankerten Grundsätzen wird auch unser weiteres Handeln in der Zukunft geleitet sein.

"Electropolis" Berlin – an Overview of 119 Years

119 years have elapsed since 1884, the founding year of the electrical age in Germany and Europe. The prime location was Berlin. It was here that the first power stations were constructed, to supply ordinary people and commercial needs. The electro-industry in Berlin provided the initial equipment for both electrical and thermal energy. In the course of an innovative interplay between technology and applied science the new power advanced by leaps and bounds in a very short space of time. It reached a range of output that we nowadays take for granted on a daily basis.
Since 1884, the Bewag and its predecessors have built and run a total of 27 power stations. On average, therefore, a new plant has been opened every four years. Moreover, five of these plants have, at least in part, often been radically reconstructed.
At the beginning of 2003, the Bewag is operating eleven up-to-the-minute thermal power stations and two "block" thermal power plants with an installed gross electrical output of some 2,750 MW, as well as a host of smaller power plants. In all these thermal power plants, the fuel-saving measure of coupling electricity and heating supplies is being implemented, which has resulted in a total heating capacity of around 6,000 MW. This feeds into the largest long-distance thermal grid in Europe, with some 1,250 kilometres of location line and more than 13,400 transmission-stations, which supplies around 600,000 flats and numerous public buildings and business premises with heating and hot water.
At every stage, there were two primary objectives for the Bewag: on the one hand, to implement the most innovative technologies and initiatives and to follow a real commitment to all the power stations as well as to their equipping; and, on the other, to supply the metropolis of Berlin with energy and heat in the most efficient manner. These principles, firmly anchored in the history of the "electropolis" Berlin, will also guide our future actions.

DAS ARCHITEKTONISCHE UND TECHNIKHISTORISCHE ERBE
THE ARCHITECTURAL AND TECHNICAL LEGACY

Die Bau- und Entwicklungsgeschichte der Berliner Elektrizitätswerke
The History of the Construction and Development of the Berliner Elektrizitätswerke

Hubert Staroste

Das heutige Denkmalprofil der Stadt ist entscheidend geprägt durch den Aufstieg Berlins zur Industriemetropole im ausgehenden 19. und beginnenden 20. Jahrhundert. Die überragende Stellung, die der Maschinenbau und die Metallindustrie in den letzten Jahrzehnten des 19. Jahrhunderts erlangten, haben den schnellen Aufstieg der Elektroindustrie in Berlin wesentlich befördert. Berlin war um 1900 vor allem eine Stadt der Elektroindustrie und der Elektrizitätswirtschaft. Die starke Konzentration von Unternehmen dieses innovativen Wirtschaftszweiges und deren dynamische Entwicklung haben Berlins Ruf als „Elektropolis" begründet und über viele Jahrzehnte getragen. Insbesondere die in Berlin ansässigen internationalen Branchenführer in der Kraftwerkstechnik, die AEG und die Firma Siemens, haben hier entscheidende Beiträge geleistet. Einerseits nahmen beide Unternehmen in den Anfangsjahren des Kraftwerksbaus häufig Prototypen in Berlin in Betrieb, die nach bestandener Bewährungsprobe weltweit zum Einsatz kamen. Andererseits gab es einen schnell wachsenden Bedarf an Licht- und Kraftstrom, den die sich dynamisch entwickelnde Industriestadt Berlin mit ihren modernen Großindustrien abforderte und damit den schnellen Übergang zum Großkraftwerksbau geradezu erzwang.

Der Bau des ersten europäischen Drehstromkraftwerkes durch die AEG in den Jahren 1895 und 1896 in Oberspree war in erster Linie den energetischen Bedürfnissen des Unternehmens selbst geschuldet: Die AEG setzte bei der maschinentechnischen Bestückung des in Oberspree angesiedelten Kabelwerkes nur Elektromotoren als Antriebstechnik ein, die ebenfalls von der AEG gefertigt wurden, zum erheblichen Teil auch als Einzelantrieb für den hochmodernen Maschinenpark, wie beispielsweise im Kupferwalzwerk. Mit dem breiten Einsatz dieser innovativen Antriebstechnik, die den industriellen Fertigungsprozess revolutionieren sollte, erlangte das Unternehmen längere Zeit einen erheblichen Vorsprung gegenüber der Konkurrenz.

The city's current high profile in terms of historic buildings has been crucially influenced by Berlin's rise to an industrial metropolis in the late 19th and early 20th centuries. The dominating position that machinery manufacture and the metal industry achieved in the closing decades of the 19th century greatly encouraged the rapid growth of the electro-industry in Berlin. In 1900, Berlin was first and foremost a city for the electro-industry and the electro-economy. The concentration of enterprises specialising in this new branch of the economy – and its dynamic development – established and carried forward for several decades Berlin's reputation as the "electropolis". Principally, it was those international leaders of the electrical industry with HQs in Berlin – such as AEG and Siemens – that played a decisive role here. On the one hand, in the early years of power station construction, both concerns often put prototypes into operation in Berlin. After successfully completing a trial period, these prototypes went into service all over the world. On the other hand, a rapidly increasing need for light and power, required by the dynamically developing industrial city of Berlin, with its up-to-the-minute large-scale industries, demanded and thus almost compelled the city to start building high-capacity power plants without delay.

The construction by the AEG of Europe's first alternating current (a.c.) power station in 1895–96 in the Oberspree district, was largely due to the enterprise's own energy requirements. In terms of providing the Oberspree power plant with technical equipment for its machinery, the AEG supplied only electric motors in the form of propulsion technology; this was also manufactured by the AEG predominantly also as the sole means of propelling the highly modern machine park – as, for instance, in the copper rolling mill. Thanks to the broad acceptance of this innovative method of propulsion which was to revolutionise the latter stages of industrial manufacture, the concern gained a considerable advantage for some time over its competitors. The electricity current

Kraftwerk Oberspree,
historische Ansicht um 1905
Oberspree Power Station,
historical view around 1905

Der für den Betrieb der Elektromotoren benötigte Kraftstrom konnte nur durch den Bau eines neuen und vor allem hochleistungsfähigen Drehstrom-Kraftwerkes verfügbar gemacht werden. Diese unternehmerischen Erfordernisse, gepaart mit dem vom AEG-Ingenieur und späteren Chefkonstrukteur Michael von Dolivo-Dobrowolski maßgeblich entwickelten Know-how, waren die entscheidenden Voraussetzungen für den Bau des ersten Drehstrom-Kraftwerkes am Standort Berlin. Die Kraftwerksarchitektur war in dieser Zeit stark durch die Produzenten beeinflusst. Sie bestimmten mit der technischen Anlage zwangsläufig auch die Grunddisposition der baulichen Anlagen, in den ersten Jahren lieferten sie die bauliche Hülle häufig gleich mit. Da es sich um eine neue Bauaufgabe handelte und man nicht auf unmittelbare Vorbilder zurückgreifen konnte, wurden für die Kraftwerkstechnik nicht selten adäquate architektonische Lösungen gefunden. Vor allem mit dem Übergang zum Bau leistungsfähiger Drehstromkraftwerke im öffentlichen Stadtraum hatte deren Architektur die Unternehmen angemessen als modern und innovativ zu repräsentieren. Die Auftraggeber aus der Elektroindustrie und der Elektrizitätswirtschaft agierten so objektiv als Förderer moderner Strömungen in der Industriearchitektur. Zumindest fällt es auf, dass in dieser Branche Architekten ein Betätigungsfeld fanden, die wie Franz Schwechten, Peter Behrens, Hans Müller oder Hans Hertlein zu den Wegbereitern oder namhaften Vertretern einer modernen Industriearchitektur gehörten. Die Berliner Kraftwerke zählen nicht zuletzt deshalb zu den eindrucksvollsten und städtebaulich markantesten Zeugnissen der „Elektropolis" Berlin. Es handelt sich hier in vielen Fällen auch um Schlüsselzeugnisse im überregionalen Vergleich, um architektur- und technikgeschichtliche Marksteine.

An den denkmalgeschützten Kraftwerksbauten und Anlagen Berlins sind die wichtigsten Abschnitte dieser nunmehr hundertjährigen Entwicklungsgeschichte in einer anderswo nicht mehr vorzufindenden Geschlossenheit ablesbar. Die Entwicklung der Kraftwerkstechnik und des Kraftwerksbaus kann grob in vier Phasen eingeteilt werden: die Pionierphase mit den Gleichstrom-Blockkraftwerken, die Phase der Drehstromkraftwerke der Kaiserzeit, die Phase der Großkraftwerke der Weimarer Zeit und schließlich die Wiederaufbau- und Modernisierungsphase Ende der 1940er Jahre und in den 1950er Jahren.

required to drive the electric motors could only be made available through the construction of a new and – above all – high-capacity alternating current power station. These entrepreneurial demands, coupled with the exceptional know-how developed by the AEG engineer and future chief designer Michael von Dolivo-Dobrowolski, were the deciding factors behind the construction of the first a.c. power station in Berlin. At this time, power-plant architecture was strongly influenced by the producers. Along with the technical layout, they inevitably also determined the basic disposition of the installations that were to be built – in the early years they often even delivered the shell of the future construction from the very start. Since this was a totally new kind of construction and it was not possible to refer to previous models, it turned out frequently to be the case that ingenious architectural solutions were found. Primarily, when it came to constructing efficient a.c. power stations in the public domain of a large city, the architecture chosen needed to show the concerns that backed it as modern and innovative. Thus, objectively speaking, the commissioning bodies from the electro-industry and electro-economy acted as promoters of up-to-date tendencies in industrial architecture. In any case, it is very evident that in this area architects discovered a field of activity which made figures like Franz Schwechten, Peter Behrens, Hans Müller or Hans Hertlein the pathfinders or illustrious representatives of a modern form of industrial architecture. It is not only for that reason that Berlin power stations are among the most impressive and most striking in terms of urban architecture of any examples of the "electropolis" of Berlin. In many cases, we are dealing here with key models, on the basis of extra-regional comparison, in other words, with architectural and technical and historical "milestones".

The listed buildings of power plants and other installations in Berlin are the most significant chapters of what is by now a century-old history of development – in no other location can this be seen in a form that is so self-contained. Roughly speaking, the development of power station technique and construction can be divided into four phases: the pioneering phase, with the direct current (d.c.) block power plants, the alternating current (a.c.) power stations of the Imperial period, the phase of high-capacity power stations in the Weimar period, and finally, the period of reconstruction and modernisation in the late 1940s and the 1950s.

Zentralstation Mauerstraße, historische Ansicht um 1900
Central Power Station , Mauerstraße, historical view around 1900

1. Die Pionierphase von 1883-1895

Die öffentliche Stromversorgung Berlins begann 1883 mit der Inbetriebnahme des Prototyps eines Blockkraftwerkes durch Emil Rathenau und Oskar von Miller im Keller der Friedrichstraße 85. Da diese Versuchsanlage ihre Bewährungsprobe bestand, gingen zwischen 1885 und 1890 die Zentralstationen in der Markgrafenstraße, der Mauerstraße, der Spandauer Straße und am Schiffbauerdamm in Betrieb. Diese vier Blockkraftwerke versorgten mit ihren Gleichstrom erzeugenden Dynamomaschinen die Berliner Innenstadt vornehmlich mit Lichtstrom. Da die Gleichstromversorgung wegen der hohen Leitungsverluste nur einen maximalen Versorgungsradius von zwei Kilometern zuließ, mussten die Zentralstationen im „Inselbetrieb" arbeiten. Von diesen vier Kraftwerksanlagen sind auf den Grundstücken Mauerstraße 78-80 und Schiffbauerdamm 21 noch anschauliche Bauteile erhalten.

Beide Zentralstationen waren wie damals üblich als Gewerbehöfe angelegt, mit an der Straße liegenden und einen Hof einfassenden Geschäftshäusern sowie flügelartig anschließendem Kesselhaus und Maschinenhäusern (Zentralstation Schiffbauerdamm) oder mit zwei gegenüberliegenden und jeweils aus Kessel- und Maschinenhaus bestehenden Anlagen (Mauerstraße). Der nördliche und noch heute bestehende Bau in der Mauerstraße war nach dem Vorbild der ersten Berliner Zentralstation in der Markgrafenstraße ursprünglich ein zweigeschossiger Bau mit im Erdgeschoss installierten Maschinensätzen und darüber liegenden Kesselanlagen. Bereits 1893 wurde die Zentralstation Mauerstraße auf leistungsfähigere Dampfmaschinen und Kesselanlagen umgerüstet und dabei das Obergeschoss mit den Kesselanlagen entfernt, um in der gewonnenen hohen Halle die neuen, deutlich größeren und leistungsstärkeren Maschinensätze aufzustellen. Es handelt sich hier um das älteste bauliche Zeugnis der Elektrizitätswirtschaft Deutschlands. Beim Umbau des Standortes zum Abspannwerk Buchhändlerhof 1926-28 durch Hans Müller blieb die alte Maschinenhalle in ihrer räumlichen Fassung von 1893 weitgehend unversehrt und ist bis heute erhalten. Müller hatte sie als Phasenschieberhalle in den nördlichen Schalthausbau integriert. Auch auf dem Standort Schiffbauerdamm 21 blieben die in den Jahren 1889 und 1890 erbaute Kesselhalle und die in den 1920er Jahren überformte und zu

1. The Pioneering Phase: 1883 – 1895

Providing Berlin with a public power supply began in 1883, with the prototype of a block power station that was put into operation by Emil Rathenau and Oskar von Miller in the basement of 85 Friedrichstraße. As this trial installation worked, between 1885 and 1890 the central plants in Markgrafenstraße, Mauerstraße, Spandauer Straße, and the Schiffbauerdamm also went on stream. Through their direct current generating dynamo-machines, these four block power plants supplied Berlin's inner city primarily with electric light. Because of heavy losses in transmission, the direct current had a maximum range of only two kilometres, the central power stations had to provide an "island supply". Of these four power-station installations, attractive vestiges of sections of buildings are still visible on the land at 78-80 Mauerstraße and 21 Schiffbauerdamm.

As was customary at the time, both central power plants were laid out as "courtyard business premises", with business houses sited on the street and enclosing a courtyard. They also had a boiler-house and machine shops (at the Schiffbauerdamm Power Station), or two installations that lay opposite one another and comprised a boiler-house and machine shop (as at Mauerstraße). Following the model of the first Berlin central power station in Markgrafenstraße, the northern, still existing construction at Mauerstraße was originally a two-storey building with machinery installed on the ground floor and boiler-house units fitted above it. As early as 1893, the Mauerstraße Central Power Station was switched to more productive steam-powered machines and boiler-house installations. This meant that the first floor, where the boiler-houses first stood, could be removed, and the high space gained in that process meant that the new machinery – much larger and more productive than the previous – could be installed. Here, we are talking about the oldest material evidence of Germany's electricity economy. In converting this plant into the Buchhändlerhof Transformer Station between 1926 and 1928, Hans Müller largely kept the old machine shop in its former proportions, dating back to 1893, and it has survived to this day. Müller integrated it into the northern transmission centre as a phase-

einem Geschossbau umgebaute Maschinenhalle erhalten. Vor allem die Kesselhalle vermittelt mit ihrer vergleichsweise gut überlieferten Baugestalt ein authentisches Bild von der Konstruktion und der Architektur dieser Blockkraftwerke.

2. Die Drehstromkraftwerke der Kaiserzeit

Waren die auf den Blockinnenflächen erbauten Blockkraftwerke noch vergleichsweise einfache Funktionsbauten mit einem bescheidenen Repräsentationsanspruch, so änderte sich dies mit dem Bau der ersten Drehstromkraftwerke grundlegend. Mit der vom AEG-Ingenieur Michael von Dolivo-Dobrowolski maßgeblich mitentwickelten Drehstromtechnik konnte elektrische Energie über größere Distanzen verlustarm zum Verbraucher geleitet werden. Die neuen Drehstromkraftwerke standen deshalb an der Peripherie der Stadt auf ausbaufähigen Arealen mit guter Verkehrsanbindung und sind mit deutlich höheren Leistungen und Bauvolumina bereits Zeugnisse der Großindustrie. Durch die Wasserlage erfolgte der Kohlentransport per Schiff, der Kohlenumschlag war bereits mechanisiert. Das Kraftwerk Oberspree erreichte bereits 1903 mit produzierten 16,7 MW zwei Drittel der Stromkapazität der fünf innerstädtischen Blockkraftwerke, die im Jahr 1898 eine Gesamtleistung von 24,836 MW hatten. Auch in Konstruktion und Architektur der 1895/96 errichteten Bauten des Drehstrom-Kraftwerkes Oberspree wird der großindustrielle Charakter der Anlage bereits deutlich. Der Hausarchitekt der AEG, Paul Tropp, war für die Entwürfe des aus Maschinen- und Kesselhaus bestehenden ältesten Teils des Kraftwerkes verantwortlich. Beide Hallen bestanden aus einer mit gelben Klinkern verkleideten Stahlskelettkonstruktion, filigrane Eisenfachwerkbinder tragen die flachen Satteldächer. Vor allem die an der Wilhelminenhofstraße giebelständig platzierte Maschinenhalle erhielt eine die moderne Konstruktion verbergende und den Bau mit reicher Bauzier „aufwertende" historistische Architektur. Diese ursprüngliche Gestaltung ist an der hofseitigen Giebelfront fast störungsfrei erhalten geblieben.

Das im Oktober 1897 in Betrieb genommene Kraftwerk wurde mehrfach erheblich erweitert, erstmals schon 1899, zwischen 1905 und 1907 dann um eine lang gestreckte, traufseitig zur Wilhelminenhofstraße angelegte dreischiffige Turbinenhalle und das parallel angefügte Kesselhaus. Mit der parallelen Anordnung von Turbinen- und Kesselhalle war eine technologisch optimale Verknüpfung von Kesselanlagen und Wärmekraftmaschinen erreicht. Zudem konnte die Stromerzeugung durch Verlängerung der Hallen und neu aufgestellte Maschinensätze problemlos erweitert werden. Nach diesem Vorbild wurden alle weiteren Berliner Drehstromkraftwerke als Parallelanlagen errichtet.

Die AEG hatte 1904 die Curtis-Patente gekauft und ab 1905 in der eigenen Turbinenfabrik leistungsfähige Dampfturbinen hergestellt. In der Turbinenhalle des Kraftwerkes Oberspree wurden 1905 erstmalig in Deutschland hochmoderne Dampfturbinen installiert. Da die Dampfturbinen bei einer doppelt so hohen Leistung nur etwa ein Viertel des Platzbedarfs der Dampfkolbenmaschinen hatten, konnten die Strom-

displacement shop. At the 21 Schiffbauerdamm site, the boiler-house installations, constructed in 1889 and 1890, as well as the machine shop, which was restructured and converted into a two-storey building in the 1920s, still exist. The boiler-house installation, with its relatively well-preserved construction, gives an especially authentic impression of the building and architecture of these block power plants.

2. The Alternating Current Power Stations of the Imperial Period

If the block power stations built on the sites inside the blocks were relatively simple and functional constructions with only a modest claim to being imposing, this changed radically with the construction of the first alternating current power plants. Thanks to alternating current technology, in whose development the AEG engineer Michael von Dolivo-Dobrowolski played a leading role, electrical energy could be supplied to the consumer over long distances with very little loss of power. Accordingly, the new a.c. power plants were built in locations near the city's periphery on sites capable of expansion with good transport links; with their markedly higher performance and greater structural volume, these are early signs of large-scale industry. The nearness of water meant that coal could be transported by ship – coal-handling was already mechanised. As early as 1903, with an output of 16.7 MW, Oberspree Power Station was producing two thirds of the power capacity supplied by the five inner-city block power plants, which, in 1895–96, had a combined total output of 24,836 MW. In terms of construction and architecture, the buildings of Oberspree A.C. Power Station – constructed in 1895/96 – also reveal the whole installation's large-scale industrial character. Paul Tropp, the resident AEG architect, was responsible for the designs of the oldest sections of this power station – consisting of machine shop and boiler-house. Both halls had a yellow-clinker covering over a steel skeleton construction, with the low saddle-back roofs supported by filigree iron struts. Especially the Wilhelminenhofstraße site, with its machine shop that boasted a gable, was given ornate "structural decoration" which concealed the modern construction and provided the building with "embellishing" historic architectural touches. On the gable-front facing the courtyard these original structures have been retained almost without any alteration.

This power station, which started operations in October, 1997, was frequently much extended – for the first time as early as 1899, then between 1905 and 1907, and finally by a lengthy, three-nave turbine hall, at right angles to Wilhelminenhofstraße, including the boiler-house that was added and ran parallel to it. The parallel installation of the turbine hall and boiler-house turned out to be a technologically ideal combination of the boiler-house units with the thermal energy machines. In addition, the output of electricity could be straightforwardly increased by the lengthening of the halls and the newly installed sets of machinery. Following this model, all future a.c. power stations in Berlin were constructed as parallel installations.

In 1904, the AEG had bought the Curtis patents, and from that date had been manufacturing efficient steam-turbines in its own turbine-factory.

Kraftwerk Oberspree, Turbinenhalle, Aufnahme um 1990
Oberspree Power Station, turbine hall, photograph around 1990

erzeugungskapazitäten im Kraftwerk Oberspree deutlich gesteigert werden. Die Zahlen weisen für das Jahr 1906 eine Gesamtleistung von 30,4 MW auf und für 1916 bereits eine Jahresleistung von 42,5 MW. Auch in ihrer Architektur zeugt die Turbinenhalle bereits von einer sich verändernden Baugesinnung: Die in eine moderne Fenster-Pfeiler-Struktur aufgelöste Fassade zeigt in der sparsam eingesetzter Bauzier neben traditionellen Formen auch symbolische Attribute der Elektrotechnik. Die Hauptfronten des alten Maschinenhauses erhielten 1912/13 eine der Turbinenhalle angeglichene und bereits von der beginnenden Moderne beeinflusste Gestaltung. Dies ist besonders prägnant an der straßenseitigen Giebelfront mit dem dominanten Segmentbogenfenster geschehen. Das Kraftwerk Oberspree wurde seit 1899 von den Berliner Elektrizitätswerken (BEW), einer gemeinsamen Tochtergesellschaft der AEG und der Stadt Berlin, betrieben und 1933 stillgelegt.
Das in den Jahren 1899 und 1900 nach Plänen von Franz Schwechten errichtete Kraftwerk Moabit (Leistung 1900: 7,2 MW, 1918: 42,4 MW) zeigte eine Architektur, die aus der Bauaufgabe heraus entwickelt wurde. Schwechten hatte bereits für die AEG-Apparatefabrik in der Ackerstraße die Fassaden entworfen und war auch noch Jahre später

Kraftwerk Moabit, historische Ansicht, um 1905
Moabit Power Station, historical view around 1905

In 1905, the turbine-hall of Oberspree Power Station saw the installation of the very latest steam-turbines for the first time in Germany. Because these steam-turbines could generate an output that was double that of the earlier steam-driven piston machines, while taking up only about a quarter of the floor area, the amount of output generated by Oberspree Power Station could be greatly increased. The figures tell their own story: in 1906 total output amounted to 30.4 MW, whilst by 1916, annual output had risen to 42.5 MW. Even in terms of its architecture, this turbine-hall bears witness to a change in construction design. The highly modern style of windows alternating with columns resulted in a more open style of façade, whilst restrained elements of architectural decoration blended in with traditional forms to suggest symbolic aspects of electro-technology. From 1912/13 the main façades of the old machine shop were given a style relating to that of the turbine-hall and which already bore the traces of modern influence. This was especially evident on the gable-façade facing the street, with its striking segmental arch windows. Oberspree Power Station had been operated by the Berliner Elektrizitätswerke (BEW) – a subsidary company of AEG and the municipality of Berlin; and it was closed down in 1933.
Moabit Power Station, built from plans by Franz Schwechten in 1899 and 1900 (output, in 1900: 7.2 MW, by 1918: 42.4 MW), displayed a form of architecture that was developed out of the construction criteria. Schwechten had previously designed the façades for the AEG appliance factory in Ackerstraße, and, years later, continued to work frequently for the AEG – amongst other projects that he completed was construction of the factory on Brunnenstraße. Responsible for the overall architectural design, Schwechten devised a magnificent form of hall-design for the machine shop in Moabit Power Station. Basing his plans on his pioneering station-architecture at Anhalter Railway Station in Berlin from 1876 to 1880, Schwechten made the steel support-girders of his machine shop into the dominant motif of his architectural design. From this basis, he created a successful symbiosis of engineering construction and architectural art. Schwechten largely eschewed the usual forms of decoration for new buildings, and concentrated instead on the restrained symbolic decorative touches which underlined the specific function of Oberspree Power Station. The "curving tail gable" on the

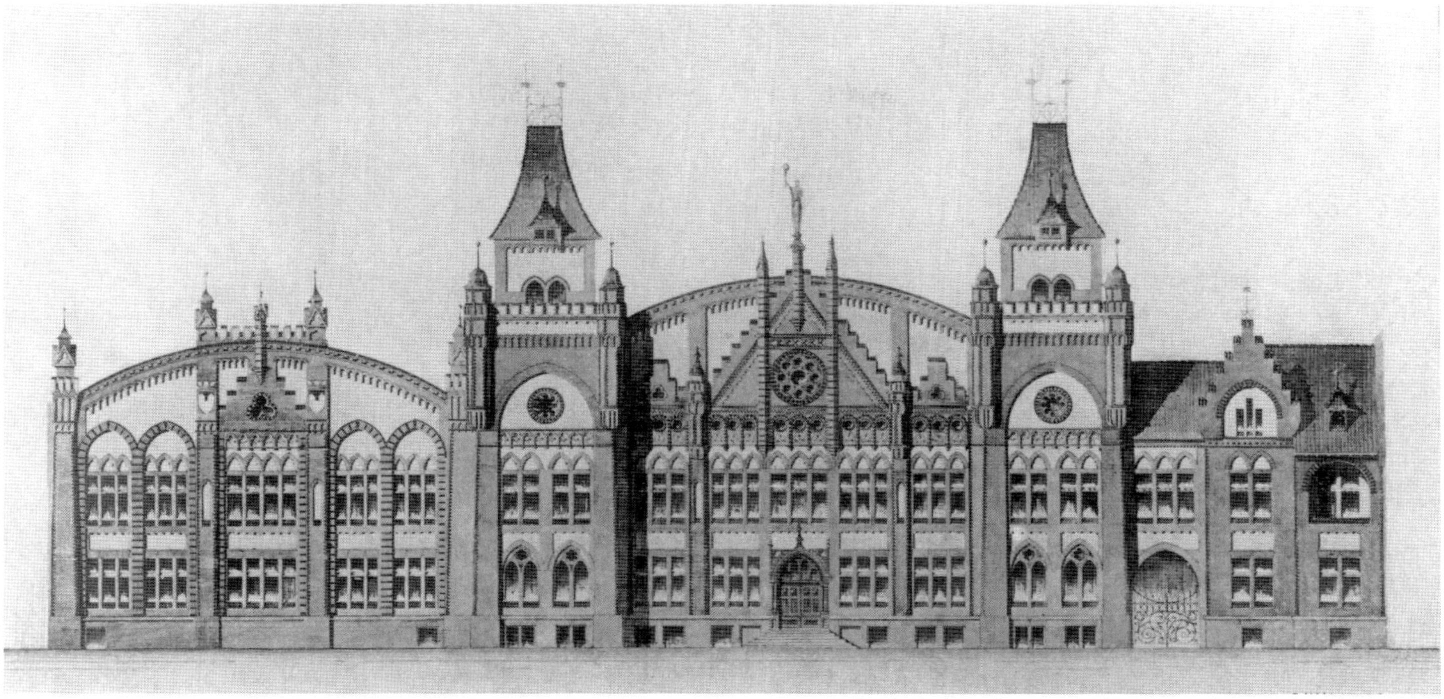

Kraftwerk Charlottenburg, Entwurfszeichnung der Hauptfassade, um 1899
Charlottenburg Power Station, design drawing of the main façade, around 1899

unter anderem beim Bau der Fabrik an der Brunnenstraße immer wieder für die AEG tätig. Für den architektonischen Gesamtentwurf verantwortlich, entwarf Schwechten für die Maschinenhalle des Kraftwerkes Moabit eine großartige Hallenarchitektur. Dem Vorbild seiner wegweisenden Bahnhofsarchitektur am Anhalter Bahnhof in Berlin aus den Jahren von 1876 bis 1880 folgend, machte Schwechten die stählerne Binderkonstruktion der Maschinenhalle zum beherrschenden Motiv seines architektonischen Entwurfs. Er schuf damit eine gelungene Symbiose von Ingenieurkonstruktion und Baukunst. Schwechten verzichtete weitgehend auf die übliche Bauzier, um noch ausgeprägter als beim Kraftwerk Oberspree mit sparsam eingesetztem symbolischem Bauschmuck auf die Zweckbestimmung des Kraftwerkes hinzuweisen. Der fast schmucklose spreeseitige Knickschweifgiebel wurde vor allem durch das riesige Rundbogenfenster dominiert. Allein das anschließende Turmmotiv wirkte als Zitat märkischer Backsteinbautradition eher gestalterisch als funktional motiviert.
Der ebenfalls von Schwechten entworfene und bereits mit Dampfturbinen bestückte Erweiterungsbau von 1907/08, Moabit II, zeigte dann bereits eine ausgeprägtere raum- und bauökonomische Entwurfshaltung, wie sie fortan den Kraftwerksbau bestimmen sollte. Im Vergleich dazu fehlte der Architektur der Maschinenhalle des 1899/1900, also zeitgleich errichteten Kraftwerkes Charlottenburg diese beeindruckende Großartigkeit. Zwar wurde auch hier die Segmentbogenform der die Halle stützenfrei überspannenden Dachbinder in der Giebelform sichtbar, sie wurde aber durch die beiden Ecktürme und die gotisierende Giebelbekrönung kaschiert. Der Giebel wirkte erst durch Nachkriegsbereinigung, der die beiden Eckturmhelme und die Giebelbekrönung zum Opfer fielen, deutlich markanter. Die Spreefront des Kraftwerkes vermittelte jedoch einen recht plastischen Eindruck von der aus Wohn- und Verwaltungsbau und der von Ecktürmen flankierten Maschinenhalle bestehenden Ursprungsbebauung. Die Projektleitung für diesen ältesten Kraftwerksteil hatte mit Georg Klingenberg ein AEG-Techniker, der die Entwicklung zum modernen Großkraftwerksbau in Deutschland und Europa maßgeblich beeinflusst und vorangebracht hat. Mit dem anschließenden Backsteinkubus des Kesselhausneubaus von 1953 und dem sachlich-expressionistischen Schalthaus von 1926 ist an der wasserseitigen Schaufront die ganze Entwicklungsgeschichte des

side facing the Spree – which had almost no ornamentation – was dominated by an enormous semi-circular arch. Only the adjoining tower motif seemed motivated more by architectural than functional concerns, quoting as it did the brickwork tradition of the Brandenburg March.
The 1907/08 extension: Moabit II, also designed by Schwechten, and already provided with steam-turbines, revealed a more pronounced instinct to save space and construction-expense money, such as was to characterise future power-plant building. That being said, this power station lacked the impressive architectural magnificence of the machine shop built for Charlottenburg Power Station – also in 1899/1900 – in other words, at the very same time. It is true that here, too, the segmental arch form of the roof-trusses spanning the hall without supports are visible in the gable-form; on the other hand, they were concealed by the two corner-towers and the Gothic-style gable-crown. Only as a result of cleaning after the War, to which both corner-tower projections and the gable-crown were subjected, did the gable appear much more striking. Even so, the power station façade that faced the Spree conveyed a truly three-dimensional impression of the original building – now only ruins – of the machine shop that was made up of accommodation and administrative quarters, flanked by a pair of corner towers. Responsibility for the project that focused on the power plant's oldest section was taken by AEG technician Georg Klingenberg who profoundly influenced and promoted the development toward high-capacity power stations in Germany and Europe. The adjoining brick cube of the new boiler-house, constructed in 1953, and the objectively expressionistic switching station of 1926 encapsulates, on the main façade facing the river – with a density that is truly remarkable – the whole history of how power plants developed. The two 1929 peak-demand turbine sets placed in the rear section of the frequently extended machine shop together with the battery connected to it in the north, comprising the 16 steel cylinders of Ruth's accumulator unit, still remain in Charlottenburg Power Station as unique evidence of power-plant technology.
Because Oberspree Power Station could not be extended any further, a new power plant, near this site, was constructed on a location near the

Kraftwerk Charlottenburg, Turbine von 1929, Aufnahme 2002
Charlottenburg Power Station, 1929 turbine, photograph 2002

Kraftwerk Rummelsburg, historische Ansicht, Aufnahme um 1910
Rummelsburg Power Station, historical view, photograph around 1910

Kraftwerkes in einer bemerkenswerten Dichte ablesbar. Mit den im hinteren Teil der mehrfach verlängerten Maschinenhalle aufgestellten beiden Spitzenlastturbinensätze von 1929 und der nördlich anschließenden, aus 16 Stahlzylindern bestehenden Batterie der Ruths'schen Speicheranlage sind im Kraftwerk Charlottenburg einmalige Zeugnisse der Kraftwerkstechnik erhalten.

Da das Kraftwerk Oberspree nicht mehr erweiterungsfähig war, wurde für das östliche Versorgungsgebiet unweit dieses Standortes ein neues Kraftwerk auf einem Spreegrundstück an der Nordgrenze Oberschöneweides errichtet. Das Kraftwerk Rummelsburg entstand in einer extrem kurzen Bauzeit von nur zehn Monaten. Ende 1907 ging es als erstes, ausschließlich nur mit hochmodernen Dampfturbinen bestücktes Kraftwerk mit 13,5 MW ans Netz. 1913 hatte es eine Jahresleistung von 20 MW, im Jahr 1918 eine Leistung von 36 MW erreicht. Der Kohlenumschlag wie auch die Beschickung der Kesselanlagen waren bereits mechanisiert. Entworfen wurde das Kraftwerk ab 1906 vom Baubüro der BEW. Das Kraftwerk Rummelsburg ist technisch wie auch architektonisch eine modernisierte Version des Kraftwerkes Oberspree in Oberschöneweide. Von 1925 bis 1926 wurde das Kesselhaus und in den Jahren von 1926 bis 1929 das Maschinenhaus erweitert und die Kraftwerksleistung auf 79,6 MW erhöht. Bei den eindrucksvollen Erweiterungsbauten, die dem Kraftwerk nach Süden ein völlig neues Gesicht gaben, handelte es sich um die noch heute bestehende expressionistische Backsteinarchitektur des Architekten Hans Müller, der vor allem in den 1920er Jahren mit seinen Bauten für die Bewag zu den führenden Berliner Industriearchitekten gehörte. Die Hallenbauten strahlen absichtsvoll etwas geheimnisvolles, archetypisches aus, eine Stimmung, die eigentlich nur Bauten erzeugen, die der Zeit getrotzt haben. Hierin lag eine Hauptabsicht der expressionistischen Architektur, und Müller ist es im höchsten Maße gelungen, eine Stimmung zu erzeugen, die über den profanen Zweck weit hinausführt und die Architektur veredelt. Die Müller'sche Erweiterung gehört zweifellos zu den eindrucksvollsten Inszenierungen Berliner Industriearchitektur.

Das 1910/11 nach Plänen von Martin Rehmer (technische Planung) und Hans Müller erbaute Kraftwerk Steglitz zählt zweifellos zu den architektonisch anspruchsvollsten Gemeindekraftwerken Deutschlands. Obwohl in den folgenden Jahrzehnten mehrfach modernisiert und 1926 zum Heizwerk umgebaut, ist die bauliche Hülle der aus

Spree at the northern city limit of Oberschöneweide. Rummelsburg Power Station was built in the extremely short time of a mere ten months. In late 1907, it was connected to the grid as the first power station to be exclusively provided with the most modern of steam-turbines, generating 13.5 MW. In 1913, it had achieved an annual output of 20 MW, and by 1918, 36 MW per year. The coal-loading and handling equipment were both already mechanised. The power station was designed from 1906 onward by the BEW building office. Rummelsburg Power Station is a technically and architecturally modernised version of Oberspree Power Station in Oberschöneweide. From 1925–1926 the boiler-house, and from 1926–1929, the machine shop were extended, and the power plant's output was increased to 79.6 MW. The impressive extensions, which gave the power plant a completely new southern aspect, featured the still existing expressionistic brickwork architecture by the architect Hans Müller, who, especially in the 1920s as a result of his Bewag buildings, became one of Berlin's leading industrial architects. His hall constructions intentionally radiate a mysterious, archetypal quality – an atmosphere characteristic only of buildings that have defied time. This was one of the main purposes of Expressionist architecture, and Müller succeeded brilliantly in evoking an atmosphere which goes far beyond a merely secular agenda, ennobling the architecture.

Steglitz Power Station, constructed in 1910/11 from plans by Martin Rehmer (technical planning) and Hans Müller, is undoubtedly one of the architecturally most ambitious parish power plants in Germany. Although it was frequently modified in succeeding decades, and converted into a thermal power station in 1926, the building's shell, consisting of the director's house, with an adjoining office wing, a machine shop and boiler-house,a switching-house and a workshop, has survived more or less intact. This complex of buildings which dominates the Teltow Canal with its red brick façades, and, above all, with its high hall-gables, is amongst the first Steglitz power plants and Hans Müller's earliest work. The virtuosely handled detail already speaks the language of industrial architecture of the early Modernist period, and gives the power plant construction an objectively imposing character. With its evocative formal style, which owes a lot to Neoclassicism, this is one of the most important Berlin industrial buildings of early Modernism.

Kraftwerk Steglitz, historische Ansicht, Aufnahme um 1912
Steglitz Power Station, historical view, photograph around 1912

Direktorenhaus, anschließendem Büroflügel, Maschinen- und Kesselhaus, Schalthaus und Werkstatt bestehenden alten Kraftwerksanlage weitgehend unverändert erhalten. Der mit seinen roten Backsteinfassaden, besonders aber mit seinem hohen Hallengiebel das Ufer des Teltowkanals dominierende Baukomplex gehört zu den frühen Steglitzer Versorgungsbauten und zum Frühwerk Hans Müllers. Die virtuos gehandhabte Detailsprache zeugt bereits vom Vokabular der Industriearchitektur der beginnenden Moderne und gibt dem Kraftwerksbau einen sachlich-repräsentativen Charakter. Das Kraftwerk Steglitz gehört mit seiner prägnanten, dem Neoklassizismus verpflichteten Formensprache zu den wichtigsten Berliner Industriebauten der beginnenden Moderne.

3. Die Großkraftwerke der Weimarer Zeit

In den 1920er und 1930er Jahren erlangte die Stromversorgung eine neue Qualität in Berlin. Dies sind die Jahrzehnte der elektrotechnischen Revolution, die die industriellen Produktionsbedingungen und den Verkehr grundlegend veränderten. Die Elektrifizierung von Industrie und Verkehr, in Berlin insbesondere die Elektrifizierung der S-Bahn und der Ausbau der U-Bahn, führte zu gewaltigen Schüben im Ausbau der Stromerzeugung und Versorgung. Mit der Bildung von Groß-Berlin 1920 verdoppelte sich die Einwohnerzahl und vervielfachte sich das Stadtgebiet. Die Elektrizität veränderte das Alltagsleben der Menschen: Waren 1914 in Berlin fünf Prozent aller privaten Haushalte an das elektrische Versorgungsnetz angeschlossen, so stieg deren Anteil 1925 auf 25 Prozent und erreichte 1938 mit 92 Prozent aller Berliner Haushalte faktisch eine flächendeckende Versorgung. Die in Berlin aufgebauten Kapazitäten – im Jahr 1918 waren es 223,5 MW – reichten bald nicht mehr aus. Bereits während des Ersten Weltkrieges war es in der Berliner Rüstungsindustrie zu Versorgungsengpässen gekommen. Abhilfe versprach die Fernstromversorgung durch Großkraftwerke, die vornehmlich direkt in den Kohlefördergebieten Strom erzeugen sollten.

Das Konzept für den Aufbau einer in Großkraftwerken zentralisierten Stromversorgung in Deutschland ist eng mit dem Techniker und AEG-Manager Georg Klingenberg verbunden. Im AEG-Vorstand für den Bau und den Betrieb von Kraftwerken zuständig, sah sein nationales Versorgungskonzept den Bau von 30 Großkraftwerken vor, die, über Hochspannungsleitungen vernetzt, technische Effizienz, hohe Wirt-

3. The High-Capacity Power Stations of the Weimar Period

In the 1920s and 1930s, supplying power achieved a new quality in Berlin. These were the decades of the electro-technical revolution, which radically altered the conditions of industrial production. The electrification of industry and transport – in Berlin especially the electrification of the S-bahn – led to massive increases of power output and supply. The formation of Greater Berlin in 1920 doubled the population and multiplied by several times the area that the city occupied. Electricity transformed people's everyday lives: in 1914, around five per cent of all private households in Berlin were connected to the mains; this proportion increased to 25 per cent and by 1938 had reached 92 per cent. In other words, supply had extended more or less everywhere. Existing power supply in Berlin – in 1918, the total figure was 223.5 MW – soon proved to be insufficient. As early as World War I, the Berlin armaments industry had encountered shortages. Help was forthcoming from the long-distance power supplied by high-capacity power stations, which were primarily meant to supply power directly in the coal-providing districts.

The whole concept of building up a centralised system of power supply in high-capacity power stations in Germany is closely associated with the technician and AEG manager, Georg Klingenberg. In his capacity for constructing and operating power stations on the Board of AEG, Klingenberg envisaged the construction of 30 high-capacity power stations nationwide, which, operating on a high-voltage grid, would guarantee technical efficiency, low cost, and stable supply. The age of long-distance power supply for Berlin began as early as June 1918, when a 132 kilometre long 100 kV overland powerline supplied electricity from Golpa-Zschornewitz Power Station in the brown coal district of Bitterfeld right into the city. In 1921, a second 100 kV powerline to Niederlausitz Power Station was set up. That being said, the city decided in the 1920s – mainly for economic reasons – greatly to increase the amount of self-supplied power. The upshot was that the construction of Klingenberg and West Power Stations established technically up-to-the-minute production capacities which were unequalled anywhere in the world at that time.

Klingenberg Power Station was built by the AEG in the space of fifteen months from 1925 – 26, following the technical concept devised by

Kraftwerk Klingenberg, nach Fertigstellung 1927
Klingenberg Power Station, after completion 1927

schaftlichkeit und Versorgungssicherheit gewährleisten sollten. Die Zeit der Fernstromversorgung Berlins begann bereits im Juni 1918, als mit der Inbetriebnahme einer 132 Kilometer langen 100-kV-Überlandleitung Strom vom Kraftwerk Golpa-Zschornewitz im Bitterfelder Braunkohlerevier in die Stadt geleitet wurde. 1921 war eine zweite 100-kV-Leitung zum Niederlausitzer Kraftwerk Trattendorf fertig gestellt. Die Stadt entschied sich jedoch in den 1920er Jahren vor allem aus wirtschaftlichen Gründen, den Selbstversorgungsgrad deutlich zu erhöhen. Im Ergebnis entstanden mit den Kraftwerken Klingenberg und West technisch hochmoderne Erzeugerkapazitäten, die damals weltweit Maßstäbe setzten.

Das Kraftwerk Klingenberg wurde 1925-26 in nur fünfzehnmonatiger Bauzeit von der AEG nach dem technischen Konzept von Georg Klingenberg (seit 1927 nach ihm benannt) als erstes Berliner Großkraftwerk für die Städtischen Berliner Elektrizitätswerke errichtet. Seit 1934 gehört es der Bewag. 1927 ging das Kraftwerk mit einer Leistung von 270 MW ans Netz. Es war mit seinen drei Hauptmaschinengruppen von je 80 MW und drei Vorschaltturbinensätzen von je 10 MW sowie seinen 16 mit Steinkohlenstaub beheizten 37-atü-Kesseln die damals größte und modernste Anlage Europas. Die Gesamtanlage war vorbildlich organisiert: Die über einen Stichkanal und einen Gleisanschluss angelieferte Kohle wurde in Schüttgruben zwischengelagert und über mechanische Förderanlagen zum Kohlenmahlhaus transportiert, dort gebrochen, getrocknet und gemahlen und als Kohlenstaub in die Feuerungsanlagen der beiden Kesselhäuser geblasen. In den anschließenden Maschinen- und Schalthäusern wurde der Strom erzeugt und in das Leitungsnetz eingespeist.

Die Architektur der großräumigen Industrieanlage, beiderseits der Köpenicker Chaussee zwischen Bahngelände und Spree angelegt, ist durch die Architekten Waltar Klingenberg und Werner Issel wirkungsvoll in den Formen der Neuen Sachlichkeit mit expressionistischen Elementen bis in die einzelnen Details gestaltet worden. Fast alle Gebäude sind unter Verwendung unterschiedlicher Mauerungstechniken mit roten Klinkern verblendete oder ausgefachte Stahlskelettbauten, nur das Schalthaus ist ein reiner Mauerwerksbau. Die konsequent funktionale Architektur der rückwärtigen Bauten, bei der die Stahlkonstruktion nach außen hin unverkleidet blieb und das Kohlenmahlhaus als technisches Großaggregat erschien, bildete einen

Georg Klingenberg (and named after him from 1927 onward), as the first high-capacity power station for the Berliner Elektrizitätswerke. Since 1934, it has belonged to the Bewag. In 1927, the power plant joined the grid with an output of 270 MW. With its three groups of main machines, each generating 80 MW, and three superposed turbine sets, generating 10 MW each, as well as its 16 37-atü-boilers (atü: Atmosphären-Überdruck), fuelled by bituminous coal, this was the largest and most up-to-date installation in Europe. The whole plant was beautifully organised: the coal, which was delivered via a branch canal and railway line was temporarily placed in storage pits, then transported by mechanised rails to the coal-milling unit. Here, it was broken up, dried and milled, then blown into the boiler-houses' two fuelling units as coal-dust. Electrical power was generated in the adjoining machine shops and switching stations – then fed into the grid.

The architecture of this spacious industrial unit on either side of Köpenicker Chaussee, lying between railway lines and the river Spree, was designed down to the last detail by the architects Waltar Klingenberg and Werner Issel in a highly effective style that combined Neue Sachlichkeit (New Objectivity) with Expressionist elements. Displaying a rich variety of different techniques, almost all the buildings were faced with red clinker or given a steel-skeleton latticework pattern. Only the switching station was built purely of brick. The avowedly functional architecture of the buildings to the rear, on which steel construction remained exposed on the outside, and the coal-milling unit, which remained largely a unit on its own, was a contrast to the imposing and monumental design of the buildings fronting the Köpenicker Chaussee and the Spree. Even today, this power station is still a notable sight on the city skyline. The building's main accent was set by the eleven-storey administrative tower-block. Although strictly vertical, this structure also displays occasional touches of ornamental clinker and a stepped-back roof-crown. Some Expressionist details are to be found here – such as the triangular windows on the top floor, the broken-up roof-crown and the bands of clinker ornamentation. On the opposite side of the street, the south side, we find the lengthy 30 kV switching station, punctuated by eight towers with staircases. Both sides of the complex are connected across the street by a converted bridge, which serves as both a cable-

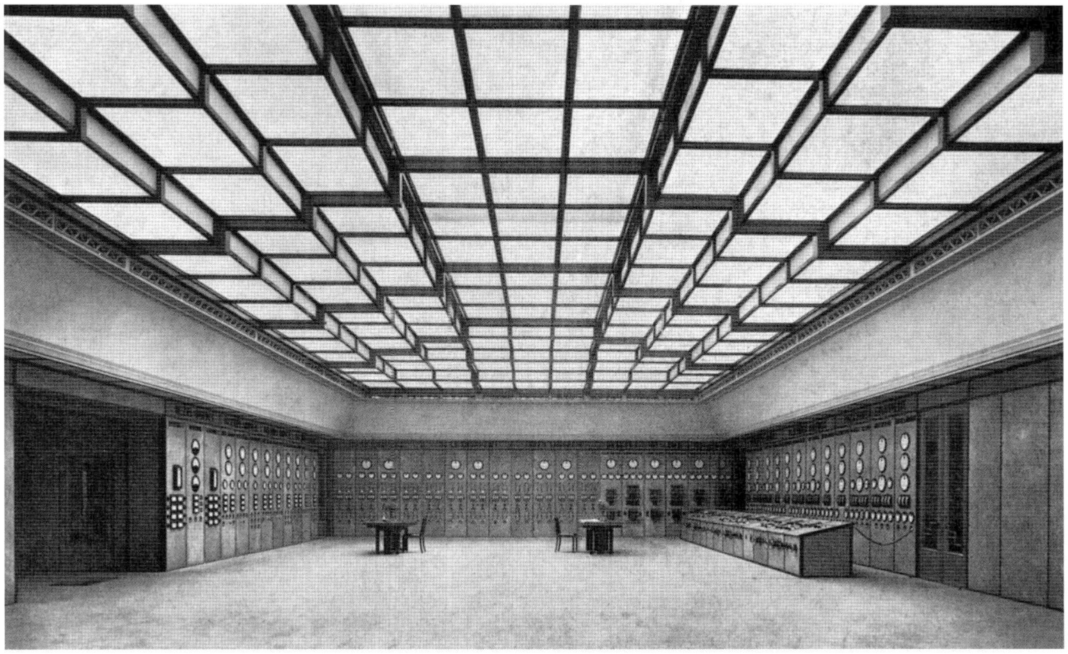

Kraftwerk Klingenberg, Schaltwarte, historische Ansicht von 1927
Klingenberg Power Station, switching observation post, historical view in 1927

Kontrast zur repräsentativen und monumentalen Gestaltung der Bauten an Köpenicker Chaussee und Spree. Das Kraftwerk ist auch heute im Stadtbild von wirkungsvoller Präsenz. Der Hauptakzent wird dabei vom elfgeschossigen Verwaltungshochhaus in strenger Vertikalisierung mit zum Teil ornamentaler Klinkerverblendung und zurückgestuftem Dachaufbau gesetzt. Hier finden sich expressionistische Details, die an den dreieckig abschließenden Fenstern im obersten Geschoss, dem durchbrochenen Dachaufsatz und den Klinker-Ornamentstreifen zum Ausdruck kommen. An der gegenüberliegenden, südlichen Straßenseite befindet sich das lang gestreckte, durch acht Treppentürme rhythmisierte 30-kV-Schalthaus. Verbunden ist der Komplex beiderseits der Straße durch eine umbaute Brücke, die als Kabeltrasse und Personalgang dient. Das technische Konzept und die Architektur der Gesamtanlage war weit über Deutschland hinaus für den Kraftwerksbau beispielgebend. In zeitgenössischen Publikationen wurde das Kraftwerk Klingenberg als ein bis in kleinste Details durchgestaltetes Gesamtkunstwerk präsentiert. 1965 wurden die die Silhouette des Kraftwerkes maßgeblich prägenden acht Stahlschornsteine durch zwei hohe Betonessen ersetzt. Mit dem 1974-87 durchgeführten Umbau zum Heizkraftwerk wurden die technischen Anlagen erneuert, die beiden Kesselhäuser durch Neubauten ersetzt und ein neuer Kraftwerksblock neben das alte Kraftwerk gestellt.

1930 ging mit der ersten Ausbaustufe das Kraftwerk West (1929-32) als zweites Berliner Großkraftwerk mit einer Gesamtleistung von 185 MW ans Netz. Ausgestattet mit neuesten Hochdruckkesseln war es damals technisch die modernste und effizienteste Anlage. Im Grundriss des Werkes spiegelt sich der technologische Prozess der Energiegewinnung wider: Er beginnt mit dem hochaufragenden Kesselhaus mit zwei Schornsteinen in Stahlbetonbauweise, an die sich das niedrigere Maschinenhaus anschließt. Der dort erzeugte Strom wird im Schalthaus auf die Verteilspannung hoch gespannt und in das Verbundnetz eingespeist. Im „Herzen" des Kraftwerks befindet sich die Warte. Dem Kesselhaus westlich vorgelagert ist die niedrigere 6-kV-Schaltanlage für den Eigenbedarf. Über Förderanlagen mit dem Kesselhaus verbunden sind der Kohlenbunker und die Entaschungsanlage am Hafen. Räumlich getrennt liegen Verwaltungs- und Lagergebäude, Werkstatt und Einlaufbauwerk. 1943 wurde das Kesselhaus zur Spree hin um das Vorschalthaus mit einem dritten Schornstein ver-

line and a walkway. The technical concept and the architecture of the whole installation turned out to be an inspiration for other power plants far beyond Germany. In contemporary publications, the Klingenberg Power Station was presented as a "total work of art" ("Gesamtkunstwerk"), carefully worked out down to the last detail. In 1965, the eight steel chimneys which characterised the power station's silhouette were replaced by two high concrete ones. In the course of rebuilding work from 1974 – 87, to convert the power station into a thermal energy plant, the technical installations were renewed, both boiler-houses being replaced by new buildings, with a new power plant block placed alongside the old power station.

In 1930, following its first extension, West Power Station (built from 1929-32) was the second Berlin high-capacity power plant – with a total output of 185 MW – to be connected to the grid. Equipped with the very latest high-pressure boiler-houses, at the time, it was the technically most up-to-date and most efficient installation anywhere. The plant's ground-plan reflects the technological process of generating energy: it starts with the towering boiler-house with two chimneys built of reinforced concrete, to which the low machine shop is joined. The power generated there is converted into high-voltage current in the switching station and stored in the distribution grid. At the "heart" of the power station we find the "observation post". Sited to the west of the boiler-house, and in front of it, is the lower 6 kV switching station for the power plant's own use. Connected to the boiler-house by transmission-lines, are the coal-bunker and the ash-removal unit by the harbour. Sited some way off are the administrative and storage buildings, as well as the workshop and delivery bay. In 1943, the boiler-house was extended toward the Spree around the superposed switching station by a third chimney. Whilst the AEG had assumed technical planning for the Klingenberg Power Station, the firm, Siemens, took responsibility for planning the construction of West Power Station. The firm of Siemens & Schuckert supplied the technical installations, and Wilhelm Dohme and Georg Trapp, from Siemens' Construction Office, were responsible for the architectural designs. In their functional "New Objectivity" style of architecture, the massive brick cubes of the power station buildings have all the markings of the "Siemens style", strikingly adopted by Hans

Kraftwerk West nach Fertigstellung 1930
West Power Station, after completion 1930

längert. Während für den Bau des Kraftwerkes Klingenberg technisch die AEG die Projektleitung übernommen hatte, lag die Projektführung beim Bau des Kraftwerkes West bei der Firma Siemens. Die technischen Anlagen lieferte die Firma Siemens&Schuckert, für die architektonischen Entwürfe waren Wilhelm Dohme und Georg Trapp aus dem Siemens-Baubüro verantwortlich. Die mächtigen Backsteinkuben der Kraftwerksbauten lassen in ihrer funktionalen und neusachlichen Architektur das Vorbild des von Hans Hertlein maßgeblich geprägten „Siemensstils" erkennen. Diese enge Verwandtschaft der neusachlichen Kraftwerksarchitektur zur neusachlichen Fabrikarchitektur der Firma Siemens ist auch Ausdruck der engen technisch-wirtschaftlichen Wechselbeziehungen von Elektroindustrie und Elektrizitätswirtschaft.

In den 1920er Jahren wurden die Kraftwerkskapazitäten jedoch nicht nur durch den Bau von neuen Kraftwerken erweitert, sondern die bestehenden Standorte wurden permanent mit modernster Technik ausgestattet und baulich ergänzt. Diese Entwicklung ist für alle Standorte zu beobachten. Es handelt sich bei den denkmalgeschützten Kraftwerken deshalb immer auch um historisch gewachsene Zeugnisse von großer Aussagekraft. So wurde das Kraftwerk Moabit zwischen 1923 und 1930 mit einer Kohlenstaubmahl- und Aufbereitungsanlage, dem 30-kV-Schalthaus, dem Umformerwerk und neuer Maschinentechnik zu einem leistungsfähigen Großkraftwerk ausgebaut. Während sich im Jahr 1918 die Leistung auf 42,4 MW belief, konnte 1925 beinahe das Doppelte, 78,9 MW, produziert werden. Vor allem mit der im Kraftwerk Moabit hier erstmalig eingesetzten Kohlenstaubfeuerung wurde Technikgeschichte geschrieben. Das 1929/30 erbaute und in seiner äußeren Form vollständig erhaltene Umformerwerk von Waltar Klingenberg und Werner Issel prägt mit seiner auch ältere Bauteile aus der Anlage Moabit II umfassenden vereinheitlichenden Gestaltung den westlichen Teil des Kraftwerkkomplexes und dokumentiert die bauliche und technische Erweiterung des Kraftwerkes in dieser Zeit. Durch den roten Backstein im Erscheinungsbild an die Schwechtenbauten angepasst, verdeutlicht das Umspannwerk den Stilwandel im Industriebau. Der schmucklose Baublock, dessen Fassaden pfeilerartig aufgelöst wirken, dokumentiert eine interessante Stilvariante sachlicher Industriearchitektur, die hier zugleich expressionistische Züge trägt.

Hertlein. This intimate relationship of "New Objectivity" power-station architecture to the matter-of-fact factory architecture typical of Siemens is also an expression of the close connection between the financially interdependent electro-industry and the electrical economy.

During the 1920s, however, power station capacities were extended not merely by the construction of new power plants. The existing sites were also equipped permanently with the very latest technology and extended. This development can be seen at all the sites considered. Thus, in the case of listed power stations, it is always a matter of historically developed signs of enormous importance. Accordingly, between 1923 and 1930 Moabit Power Station was extended into what is now a very productive high-capacity power station, with a coal-milling and preparation unit, the 30 kV switching station, the converter unit and new machinery techniques. Whereas output amounted to only 42.4 MW in 1918, by 1925, the station was generating almost double that: 78.9 MW. It was primarily thanks to the new method of coal-dust firing – here introduced for the first time ever – that Moabit Power Station made technical history. Even in its older sections taken from Moabit II, the superposed plant designed by Waltar Klingenberg and Werner Issel – built from 1929/30 and still existing on the exterior – comprises the western part of the power station complex and provides documentary evidence of the power plant's construction and technical expansion at this time. Blending in with the Schwechten buildings by its choice of red brickwork, the transformer station highlights the stylistic change in industrial constructions. The unadorned building block, whose façades look as detached as pillars, document a fascinating stylistic change in "New Objectivity" industrial architecture, which also reveals Expressionist traces here.

Charlottenburg Power Station was also completely modernised between 1925 and 1927, and extended to become the first high-pressure power station with an output of 52 MW. In this context, the two switching stations were both executed in expressionistic brickwork architecture – the 30kV switching station completed in 1926 from plans drawn up by Schönburg, and the transformer station, built in 1927/28. The 125 metre high chimney that dominates the power station silhouette was erected in 1929.

30-kV-Schalthaus des Kraftwerkes Charlottenburg, historische Ansicht um 1930
30 kV switching house of Charlottenburg Power Station, historical view around 1930

Auch das Kraftwerk Charlottenburg wurde in den Jahren von 1925 bis 1927 total modernisiert und zum ersten deutschen Hochdruckkraftwerk mit einer Leistung von 52 MW ausgebaut. In diesem Zusammenhang entstanden die beiden in expressionistischer Backsteinarchitektur ausgeführten Schalthäuser, das 1926 nach Plänen von Schönburg erbaute 30-kV-Schalthaus und das 1927/28 errichtete Fernabspannwerk. Der die Kraftwerkssilhouette beherrschende 125 Meter hohe Schornstein entstand 1929.

4. Die Wiederaufbau- und Modernisierungsphase vom Ende der 1940er Jahre bis in die 1950er Jahre

Die Berliner Kraftwerke hatten 1945 vor allem durch Kriegszerstörung nur noch eine nutzbare Leistung von 77 Prozent im Vergleich zu 1939 (1939: 747,5 MW, 1945: 576,9 MW). Noch viel gravierender waren jedoch die Folgen von Demontage und Reparationslasten. Die Demontage von Hauptaggregaten durch die sowjetische Besatzungsmacht betraf vor allem die Werke im Westteil der Stadt. Diese Situation wurde durch die Berlin-Blockade 1948/49 und die wirtschaftliche Teilung der Stadt noch erheblich verschärft. Eine Schlüsselrolle kam in den Anfangsjahren der „Strominsel" West-Berlin dem Wiederaufbau des Kraftwerkes West zu, das 1953 nach dem Regierenden Bürgermeister Ernst Reuter benannt wurde. Die Ausrüstungen mussten während der Blockadezeit über die Luftbrücke eingeflogen werden. Bereits ab Dezember 1949 hatte das Kraftwerk eine Leistung von 50 MW, die bis 1960 sukzessive auf 308 MW gesteigert werden konnte. Die baulichen Erweiterungen wurden auf überzeugende Weise in die bestehende Anlage integriert. Im Rahmen einer umfassenden Modernisierung entstand zwischen 1953/54 mit dem mit Backstein verkleideten, hoch aufragenden Kesselhauskubus nach Plänen von Willy Rathge ein weiteres markantes Zeugnis dieser Wiederaufbauphase im Kraftwerk Charlottenburg. Das 41 Meter hohe Kesselhaus ist Ausdruck des in moderne Kraftwerksblöcke zusammengefassten und aus je einem Hochleistungskessel und einer Kondensationsturbine bestehenden Maschinensatzes.

Mit den Kraftwerken Charlottenburg, Moabit und Steglitz wurden in den letzten Jahren traditionsreiche Standorte ganz oder teilweise stillgelegt. Diese Bauten unter Bewahrung ihrer denkmalrelevanten Substanz einer neuen Nutzung zuzuführen, ist eine anspruchsvolle sowie lohnenswerte Aufgabe.

4. The Phases of Reconstruction and Modernisation from the Late 1940s to the 1950s

In 1945 – primarily as a result of war destruction – power stations in Berlin could use only 77 per cent of their output, by comparison to 1939 (1939: 747.5 MW, 1945: 576.9 MW). Much more serious, however, were the consequences of demolition and reparation. Demolition of main complexes by the Soviet occupation forces mainly affected plants in the city's western sector. This situation was greatly worsened by the Berlin blockade of 1948/49 and the economic partition of the city. In the early years of what was called the "power island", a key role fell to West Power Station, which, in 1953, was named after the governing mayor, Ernst Reuter. During the blockade, all necessary equipment had to be flown in across the "air-bridge". By as early as 1949, the power station had a capacity of 50 MW, which increased successively until 1960 to 308 MW. Extensions to the buildings were integrated convincingly into the existing installation. In the course of a comprehensive modernisation programme, between 1953 and 1954, a further sign of this phase to rebuild Charlottenburg Power Station happened when the towering boiler-house cube – which had been clad in brick – was reconstructed from plans drawn up by Willy Rathge. The 41 metre high boiler-house is an expression of the very latest concept of modern power plant blocks and of a machine-set comprising a high-capacity boiler and a "condensation turbine".

With Charlottenburg, Moabit and Steglitz Power Stations all coming on stream, over the past few years, highly traditional sites have been completely or partially closed down. To prepare these plants for new uses, whilst retaining their listed buildings, is both an ambitious and worthwhile task that is now facing us.

Kraftwerk Reuter, Wiederaufbau nach 1945
Reuter Power Station, reconstruction after 1945

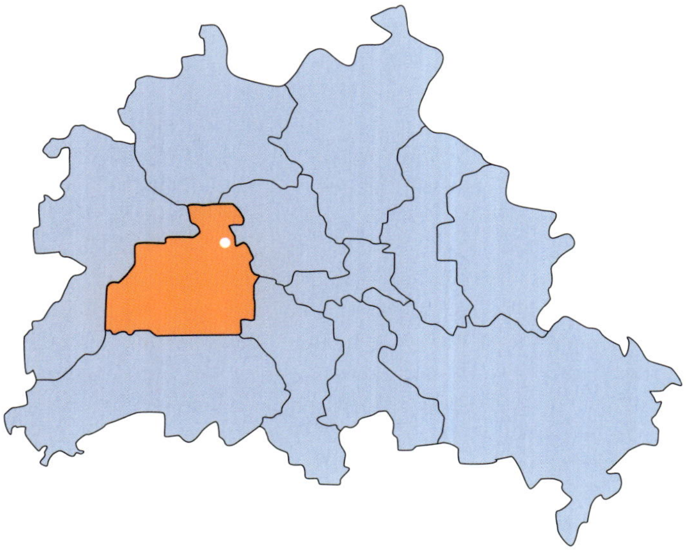

Kraftwerk Charlottenburg
Charlottenburg Power Station

Nicht mehr genutzte Anlagen der Stromerzeugung
Sites No Longer Used for Electricity Production

Christina Keseberg

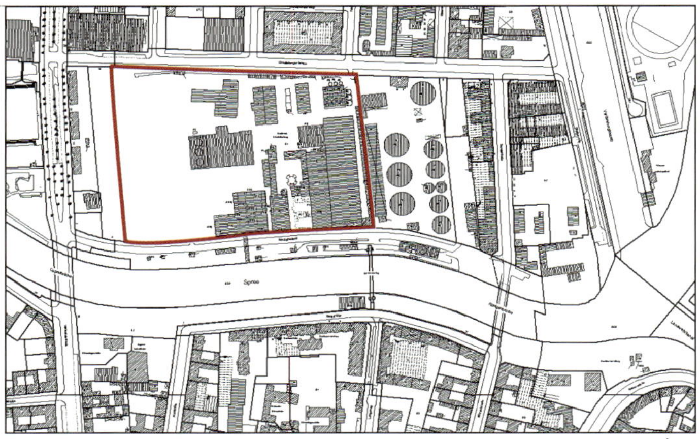

Lageplan
Site plan

DAS KRAFTWERK CHARLOTTENBURG
CHARLOTTENBURG POWER STATION

Am Spreebord 5-7

Die stadträumliche Einbindung

Das Kraftwerk Charlottenburg liegt in stadträumlich prominenter Lage am Nordufer der Spree im Zentrum Charlottenburgs. Aufgrund dieser Lage am Fluss hat man von der Schlossbrücke, der Dovebrücke und der Caprivibrücke besonders eindrucksvolle Perspektiven auf das Kraftwerk. Von der gegenüberliegenden Uferpromenade an der Iburger Straße präsentiert sich die prägnante Silhouette des Kraftwerksensembles, die auf die bisherige industrielle Nutzung des Ufers hinweist.

Südlich der Spree, vom Kraftwerksgelände über den Siemenssteig für den Fußgänger erschlossen, liegt der alte Ortskern Lietzow. Hier steht auch das Rathaus Charlottenburg. Das nördliche Umfeld weist eine Mischnutzung aus Wohnen und Gewerbe auf. Westlich, flussabwärts, liegt das Charlottenburger Schloss mit Schlossgarten. Flussaufwärts schließt sich jenseits der Gewässerkreuzung von Spree, Landwehrkanal und Charlottenburger Verbindungskanal der Spreebogen an, der durch Bauten wissenschaftlicher Einrichtungen besetzt ist.

Das Kraftwerksgelände ist verkehrstechnisch günstig erschlossen - sowohl durch die nahe gelegenen Autobahnanschlüsse und ein dichtes Netz des öffentlichen Nahverkehrs in unmittelbarer Umgebung als auch durch die Nähe zum Flughafen Tegel.

Urban Integration

The Charlottenburg Power Station is located on a prominent urban site, the north bank of the river Spree, in the centre of Berlin's Charlottenburg district. Due to its riverside situation, there are particularly impressive views of this power station from the bridges, Schlossbrücke, Dovebrücke and Caprivibrücke. From the promenade on the opposite bank, at Iburger Straße, the distinctive silhouette of the power station ensemble indicates the riverbank's previous industrial use.

To the south of the Spree, from the power station site, accessible to pedestrians by a bridge, the Siemenssteig, lies Lietzow, the district's old centre. Here, too, is Charlottenburg town hall. The outlying area to the north features a mixture of residential and commercial use. Charlottenburg castle and castle gardens are to the west, downstream. Upstream, beyond the waterway intersection of Spree, Landwehr Canal and Charlottenburg canal-link, the bend in the river Spree creates an area which is occupied by scientific institutions.

The power station site is easily accessible both by an integrated transport system, with access to nearby motorway junctions, as well as a good local transport network in the immediate vicinity and the nearby airport of Berlin-Tegel.

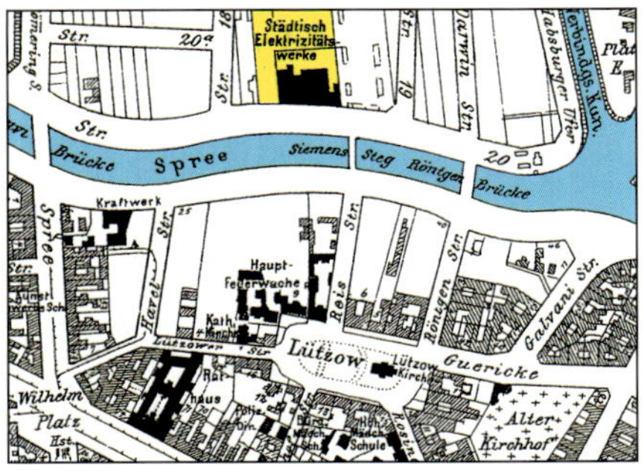

Übersichtsplan 1908 1:15000
Overview plan 1908

Entwicklungsstufen des Kraftwerks
Power station's development in phases

1900

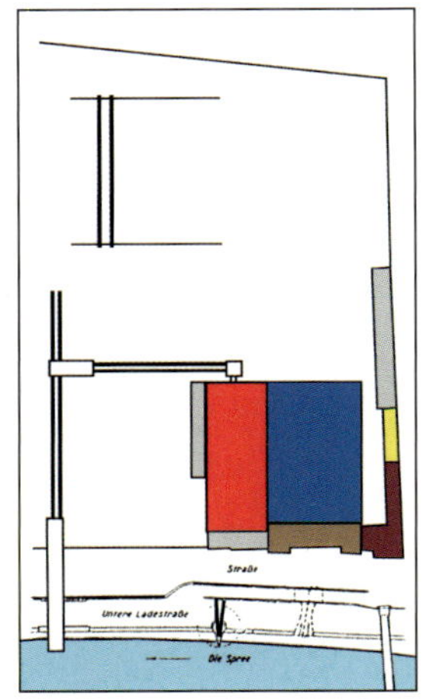
1913

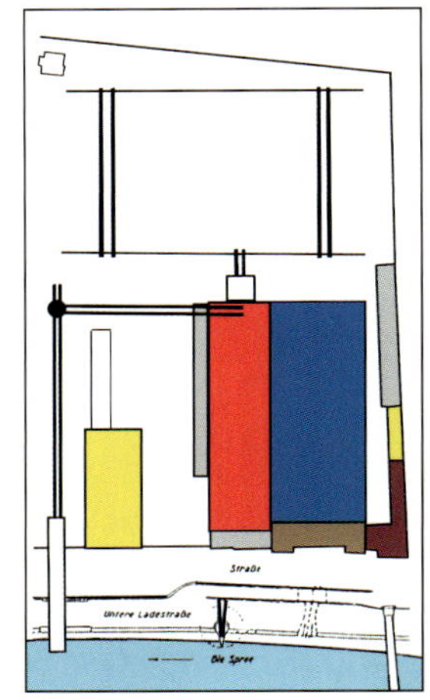
1925

Die Entwicklung des Kraftwerks

Errichtung und Ausbau bis zum Ersten Weltkrieg

Ende des 19. Jahrhunderts hatte sich die damals noch unabhängige Gemeinde Charlottenburg zu einer pulsierenden und reichen Großstadt entwickelt. Die Frage der Elektrifizierung der sich entwickelnden Industrie, der Wohnungs- und Straßenbeleuchtung und der gemeinsam mit Berlin geplanten elektrischen Straßenbahn führte 1898 zu dem Beschluss der Stadtgemeinde, ein eigenes kommunales Drehstromkraftwerk zu errichten. Das Kraftwerk wurde in den Jahren 1899/1900 auf einem Gelände im Zentrum des künftigen Entwicklungsgebietes nördlich der Spree und des bis dahin unbebauten Stadtgebietes errichtet. Geringe Grundstückspreise am Stadtrand, eine problemlose Versorgung des Kraftwerks über den Wasserweg sowie gute Bedingungen der Energieverteilung für die geplante Stadterweiterung nach Norden sprachen für den Standort.

Der Chefprojektant des Kraftwerks war Georg Klingenberg, Dozent an der Technischen Hochschule Charlottenburg und ab 1902 Vorstandsmitglied der AEG.

Kern der Anlage waren Kessel- und Maschinenhaus, zwei senkrecht zur Spree gestellte, nebeneinander liegende Hallenbauten, die an der Ostseite durch ein Wohnhaus für den Betriebsdirektor und Maschinen-

The Development of the Power Station

Building and Extension until World War One

At the end of the 19th century, the parish of Charlottenburg, at that time still independent, developed into a bustling and wealthy major city. The question of electrification for growing industry, for residential and street lighting and the electric tramcars that were planned together with Berlin, led in 1898 to the decision by the city council to construct their own, municipal alternating current power station. It was built in 1899/1900 on an inner-city site of a future development area, to the north of the Spree. This part of the city was, until that time, still empty. The low land prices on the city perimeter, the problem-free supply to the power station over the waterway and good conditions for energy distribution for the planned city expansion to the north supported this choice of location.

The chief project manager, Georg Klingenberg, was a lecturer at the Technical College in Charlottenburg and, from 1902, he was a member of AEG's board of directors.

The core of the power plant were the boiler-house and machine shop, two halls, built next to each other and vertical to the river Spree and

Historischer Grundriss 1900
Historical ground plan 1900

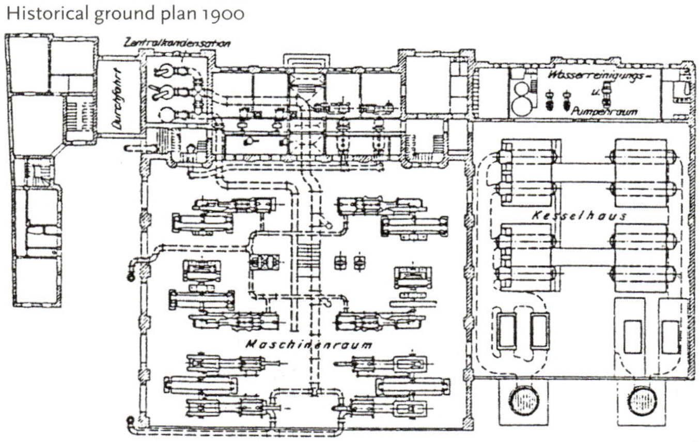

Historischer Schnitt 1900
Historical section 1900

1930

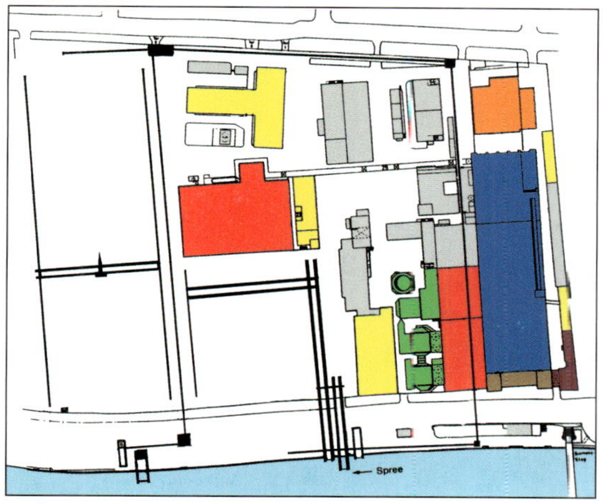

1965

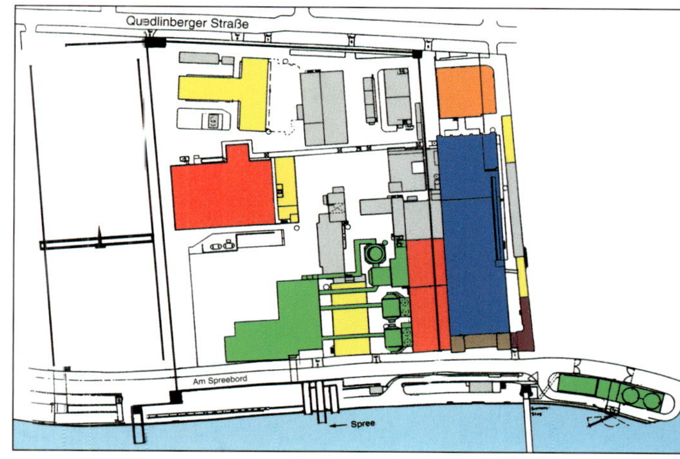

2000

Wasserstraßen	Waterways
Kesselhaus	Boiler-house
Maschinenhaus	Machine Hall
Kohlentransport	Coal Transport
Stromverteilung	Electricity Distribution
Stromspeicherung	Electricity Storage
Umweltschutz	Environmental Protection
Verwaltung	Administration
Wohnen	Accommodation
Nebengebäude aller Art	Various Outbuildings

meister ergänzt wurden. Zur gleichen Zeit wurde eine Eisenfachwerkbrücke über die Spree, der spätere Siemenssteig, errichtet.
Das Kraftwerk ging 1900 bei einer Betriebsspannung von 3.000 V mit einer Leistung von 1.500 kW in Betrieb und war als eines der ersten im Berliner Raum mit der neu entwickelten Drehstromtechnik ausgestattet.
Wegen des schnell wachsenden Bedarfs, insbesondere für den Lichtstrom, waren bereits kurz nach der Inbetriebnahme die ersten Erweiterungen notwendig. Mit dem Ausbau der technischen Kapazität wurden bis 1913 die Hallenbauten von Maschinen- und Kesselhaus auf die doppelte Größe nach Norden erweitert und zwei weitere Schornsteine errichtet. Der Seitenflügel des Wohnhauses wurde durch ein neues Bürogebäude ergänzt.
Ab 1912 lieferte das Kraftwerk zusätzlich Dampf zum Beheizen des erweiterten Rathauses Charlottenburg und leitete damit die Geschichte der Fernwärmeversorgung in Berlin ein.

Der Ausbau zum Hochdruck- und Spitzenlastkraftwerk und der Beginn der Städteheizung in den 1920er Jahren
Mit der Bildung von Groß-Berlin wurde das Werk 1922 von der Bewag übernommen und zum ersten Hochdruckkraftwerk in Deutschland ausgebaut. Um die modernen technischen Anlagen unterzubringen, mussten einschneidende bauliche Eingriffe vorgenommen werden. Das Kesselhaus wurde bis auf den Kopfbau abgerissen und nach einem

extended, on the eastern side, by accommodation for the plant's director and the chief machine engineer. At the same time, an iron-latticework bridge was built over the Spree, later known as the Siemenssteig.
This power station went on stream in 1900 with a plant capacity of 3,000 V, with an output of 1,500 kW, and it was amongst the first works in the Berlin area to be equipped with the newly developed alternating current technology. Due to the rapid increase in demand, especially for lighting electricity, soon after operations began, the first extensions were necessary. With the expansion of technical capacity, until 1913, the halls for the machine and boiler-houses were doubled in size and extended to the north; and two more chimneys were built. The side-wing of the living quarters was expanded with a new office building.
From 1912, the power station also supplied steam to heat the extension to Charlottenburg town hall, thus introducing the history of long-distance thermal energy supply in Berlin.

The Expansion to High-Pressure and Peak Demand Power Station and the Beginning of Urban Heating in the 1920s
With the creation of Greater Berlin, in 1922, the plant was taken over by the Bewag and developed into Germany's first high-pressure power station. Radical construction alterations had to be undertaken to house the modern, technical equipment. The boiler-house was pulled down, leaving only the head structure to be redesigned according to new

110-kV-Schalthaus, Errichtung 1929
110 kV Switching house, built 1929

Gesamtansicht von der Spree, Aufnahme 1951
Global view from the Spree, photograph 1951

Entwurf des Architekten Dohme neu konzipiert, das Maschinenhaus nach Norden verlängert. Mit der Umstellung der Bewag-Verteilungsspannung von 6 auf 30 kV entstand nach Plänen des Architekten Schönburg 1926 ein 30-kV-Schalthaus mit Lagergebäude. Der Lager- und Bürotrakt an der östlichen Grundstücksgrenze wurde verlängert, eine Eigenbedarfsanlage ergänzt. Kran und Förderanlagen wurden modernisiert und die alten Schornsteine abgetragen. Die Leistung des Kraftwerks betrug nach der Modernisierung 52.000 kW, der Abdampf wurde in einem der ersten Fernwärmenetze Deutschlands zur Wärmeversorgung der umliegenden Wohn- und Industriegebiete genutzt.

Ende der 20er Jahre wurde die Anlage noch einmal erweitert und zum Spitzenlastkraftwerk ausgebaut. Auf einer nordwestlichen Erweiterungsfläche an der Quedlinburger Straße entstand ein 110-kV-Abspannwerk. Der dreigeschossige, im Detail expressionistisch ausgeformte Bau mit rotvioletten Klinkerfassaden hat bis heute kaum Änderungen erfahren, die Anlage ist weiterhin in Betrieb.

Als Wahrzeichen der Kraftwerkssilhouette entstand in den Jahren 1928/29 der bis auf eine Höhe von 125 Metern geführte neue Schornstein, der seither das Stadtbild überragt.

1929 wurde zur Abdeckung von kurzzeitigen Spitzenlasten die Ruths-Speicheranlage als größte ihrer Art an die nochmals verlängerte Maschinenhalle angebunden. Die 16 erstmalig in Senkrechtstellung ausgeführten Dampfspeicher waren mit dem dazugehörigen Turbinensatz in der erweiterten Maschinenhalle bis 1995 in Betrieb.

Mit dem Ausbau Charlottenburgs zum Spitzenlastkraftwerk und einer dadurch installierten Leistung von 100.000 kW konnte der Standort mit geringen Erweiterungen in den 1930er Jahren den Anforderungen der Folgejahre bis zum Zweiten Weltkrieg gerecht werden.

Die Entwicklung nach dem Zweiten Weltkrieg und die Stilllegung des konventionellen Kraftwerksbetriebs in den Jahren von 1954 bis 2001

Im Zweiten Weltkrieg wurde das Kraftwerk nur geringfügig in Mitleidenschaft gezogen, nach Kriegsende jedoch teilweise demontiert, sodass sich die elektrische Leistung auf 63.000 kW reduzierte.

In den frühen 1950er Jahren begann mit steigendem Stromverbrauch unter den besonderen Bedingungen des „Inselnetzes" West-Berlins und dem wachsenden Interesse an der Fernwärmeversorgung die zwei-

plans, drawn up by the architect, Dohme. The machine house was extended to the north. After the conversion to the Bewag distribution-capacity, from 6 to 30 kV, a 30 kV switching house, with a storage building, was created in 1926, following plans by the architect, Schönburg. The storage and office wing on the site's eastern perimeter were lengthened and extended by an area for private use. Crane and other conveying tracks were modernised and the old chimneys were dismantled. After modernisation, the plant's capacity was 52,000 kW and waste steam was used in one of the first long-distance thermal grids in Germany to supply heat to neighbouring residential and industrial sites.

At the end of the 1920s, the plant was expanded again and developed into a peak demand power station. A 110 kV installation was built at the north-western part of the extension, at Quedlinburger Straße. The three-story structure, detailed in expressionist style with red-violet clinker façades, has to this day hardly seen any changes; and the plant continues in service.

As a symbol of the power station's sky-line, during 1928/29, a new chimney was built to a height of 125 metres, since then, it has towered above the city landscape.

In 1929, to cover short-term peak-period demand, Ruth's accumulator-unit, the biggest of its kind, was connected to the re-extended machine halls. The 16 steam accumulators, built upright for the first time, were in operation until 1995, along with the turbine set in the expanded machine shop building.

Charlottenburg's expansion to a peak demand power station and the installed capacity of 100,000 kW meant that, with few further developments in the 1930s, this site was able to cope with the demands in the years leading up to World War Two.

Post-War Development and the Closure of Conventional Power Station Operations from 1954 to 2001

In World War Two, the power station was only marginally affected. In the post-war period, however, it was partly dismantled, so that electrical capacity was reduced to 63,000 kW.

In the early 1950s, increases in energy consumption under the special conditions of the "island grid", West Berlin, and the growing interest in

Schornstein, 125m hoch, Errichtung 1928
Chimney, 125 metres heigh, built 1928

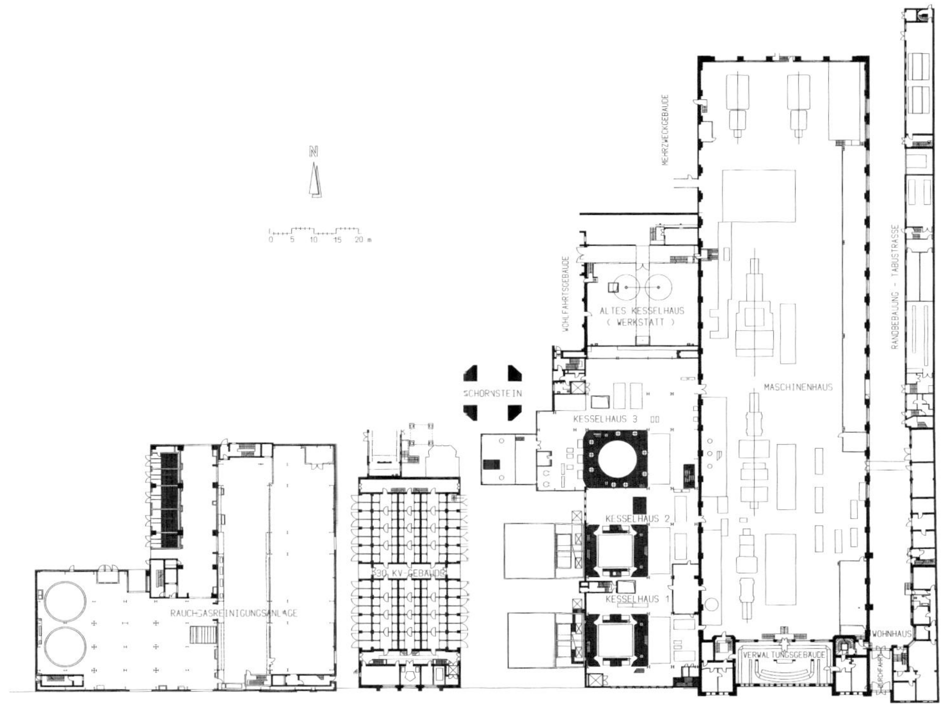

Aktueller Grundriss
Up-to-date ground plan

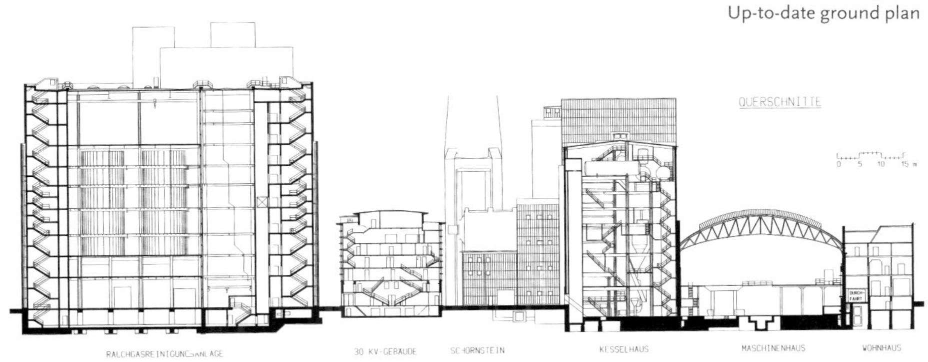

Aktueller Schnitt
Up-to-date section

te umfassende Modernisierung. In mehreren Bauetappen entstand zwischen 1954 und 1966 unter der Leitung des Bewag-Baudirektors Rathge ein gänzlich neues Kraftwerk, zusammen mit dem Aufbau eines neuen Fernheiznetzes, das in den 1960er Jahren mit dem Wärmenetz von Moabit und Reuter zusammengeschlossen werden konnte.
Die Kesselanlagen mussten völlig neu konzipiert werden. Dabei trat ein gegenüber den Altbauten hoch aufragender kubischer Hallenbau an die Stelle des alten Kesselhauses, das bis auf den aus den 1920er Jahren stammenden Nordteil abgerissen wurde. Das Rudiment wurde 1985 zu einer Werkstatt umgebaut, der alte gemauerte Schornstein blieb erhalten. Im Gegensatz zum Kesselhaus blieb die bauliche Hülle des Maschinenhauses unverändert, der maschinentechnische Teil wurde jedoch bis auf die beiden Turbinen für die Ruths-Speicher komplett erneuert.
Zur Ergänzung der Stromversorgung West-Berlins bei Lastspitzen wurde 1975 eine neue Gasturbinenanlage in Betrieb genommen. Der monumentale, annähernd regelmäßige Kubus wurde südlich des 110-kV-Schalthauses errichtet und von drei jeweils 80 Meter hohen Schornsteinen beherrscht. Die Anlage wird für die Fernwärmeversorgung und zur Deckung von Spitzenlasten in Betrieb bleiben.
Mit der dritten großen Erweiterung trug die Bewag im Rahmen eines umfassenden Kraftwerksmodernisierungs-Programmes den veränderten Anforderungen des Umweltschutzes Rechnung. Westlich des 30-kV-Schalthauses auf dem ehemaligen Kohlenlagerplatz entstand 1989 nach den Plänen von Walter Henn eine Rauchgasentschwefelungs- und Entstickungsanlage. Mit ihren Dimensionen die übrige Anlage weit

long-distance thermal energy supply led to a second phase of comprehensive modernisation. In several construction phases, from 1954 to 1966, a completely new power station was created, under the directorship of the Bewag construction chief, Rathge. That project ran alongside the development of a new long-distance thermal grid that, in the 1960s, was integrated into the thermal grid of Moabit and Reuter.
The boiler installations had to be completely redesigned. By comparison to the old structures, a cubic, steeply rising hall-structure replaced the old boiler-house that was demolished with the exception of the northern part, originating from the 1920s. In 1985, the basic structure was converted into a workshop, and the old stonework of the chimney was preserved. By contrast to the boiler-house, the residual shell of the machine shop remained unchanged. The part housing machine technology was, however, completely refurbished, with the exception of the two turbines for Ruth's accumulators.
To supplement West Berlin's electricity supply at peak periods, in 1975, a new gas turbine installation went into operation. The monumental, almost regular cube-shape was built to the south of the 110 kV switching house and dominated by three, tall chimneys, each 80 metres in height. The plant will stay in operation for long-distance thermal heating supply and to meet peak demands.
With the third, major expansion, in the context of a comprehensive programme to modernise power stations, the Bewag took into account the changed requirements for environmental protection. To the west of the 30 kV switching house, on the site of the former coal depot, a plant was

Neubau Kesselhaus 1954
Boiler-house reconstruction 1954

Gasturbine im Bau 1978
Gas turbine, under construction 1978

übertreffend hat sie die Kraftwerkssilhouette entscheidend verändert und geprägt.
Die Liberalisierung des Strommarktes brachte aus wirtschaftlichen Erwägungen das Aus für das Kraftwerk Charlottenburg als Elektrizitätslieferant. 1995 wurde die Ruths-Speicheranlage außer Betrieb genommen. Mit der Stilllegung von Block 3 endete Anfang 2001 die konventionelle Stromerzeugung im Kraftwerk. Künftig wird Charlottenburg nur noch Fernwärme liefern und bei Spitzenlasten zugeschaltet werden.

created in 1989, to plans drawn up by Walter Henn, to extract sulphur and nitrogen from flue-fumes. The scale of this plant far outweighed the other installation, decisively and characteristically changing the power station's silhouette.
Deregulation of electricity markets saw to it that, on economic grounds, Charlottenburg Power Station ceased to operate as an electricity provider. In 1995, Ruth's accumulator-unit was taken out of service. The closure of Block 3 at the beginning of 2001 put a stop to the power station's conventional electricity production. In future, Charlottenburg will only supply long-distance thermal heat and be switched on for peak period demand.

REA im Bau 1988
REA, under construction 1988

Historische Schaufassade zur Spree, Aufnahme 1925
Historical show façade, towards the Spree, photograph 1925

Die Gebäude

Das Maschinenhaus mit Verwaltungs- und Wohngebäude

Das Kraftwerk Charlottenburg ist mit seinem historischen Kern von 1899/1900 eines der beiden ältesten erhaltenen Drehstromkraftwerke Berlins und der bedeutendste Vertreter der historisierenden Werksanlagen der Kraftwerksgeneration im ausgehenden 19. und frühen 20. Jahrhundert.
Der Entwurf von Georg Klingenberg war aus einem engeren Wettbewerb hervorgegangen und entsprach offensichtlich dem Bedürfnis der aufstrebenden Stadt Charlottenburg nach einer repräsentativen, dem Selbstbewusstsein der wohlhabenden Bürgerschaft angemessenen architektonischen Gestaltung. Das Maschinenhaus ist zusammen mit dem Wohnhaus das einzige Zeugnis der ursprünglichen Anlage.
Bei Gründung des Kraftwerks war nur ein geringer Teil des zur Verfügung stehenden Geländes bebaut worden. An der Südseite des Grundstückes zum Wasser orientiert, bildeten Maschinen- und Kesselhaus mit ihren vorgelagerten Kopfbauten zusammen mit dem Wohnhaus das alte Kraftwerksensemble.
Die Schaufassade zur Spree nahm deutlich Bezug auf die Formensprache der Spätgotik, speziell auf die Tradition der märkischen Backsteingotik. Zwei kräftige Eckpylone mit dazwischen eingespannter Portalwand verkörperten das große Kompositionsmotiv und markierten die Dreiteilung des Ensembles entsprechend den dahinter verborgenen Funktionen. Das kontrastreiche Wechselspiel von rotem Backstein und hellen Putzflächen ist hier gelungen inszeniert. Zwischen den beiden Turmrisaliten mit Zinnenkranz und Helm überspielte ein gotischer Stufengiebel den zurückhaltend gestalteten Hallengiebel. Das zentrale Maschinenhaus dominierte mit seiner Ausdehnung und motivischen Überhöhung die Fassaden von Kesselhaus und Wohngebäude.
Dem hohen ästhetischen Anspruch an die Außenwirkung sollte auch das Innere der Halle entsprechen. Gestalterischer Höhepunkt war die Schaltgalerie an der Stirnwand. Hinter den drei mittleren, mit einem aufspringenden Tudor-Giebel-Dekor überfangenen Spitzbogenöffnungen wurde die Schaltanlage in ihrer zentralen Bedeutung beinahe sakral hervorgehoben. Die beiden äußeren Öffnungen vermittelten den Übergang zum Kopfbau mit dem Verwaltungstrakt. Heute sind die Bögen der Galerie geschlossen.

Maschinenhalle, Aufnahme 1907
Machine hall, photograph 1907

The Buildings

Machine Hall with Administration and Accommodation

The Charlottenburg Power Station, with its historic core dating from 1899/1900, is one of Berlin's two oldest remaining alternating current power stations and the most significant representative of the historical sites of the generation of power plants from the turn of the 19th and early 20th century.
Georg Klingenberg's design was the winner of a close competition and evidently matched the need of the up-and-coming city of Charlottenburg for a representative, architecturally imposing structure that reflected the self-image of a wealthy citizenship. The machine shop, along with the living accommodation is the only remaining part of the original site.
When the power station was founded, only a minor part of the available land was used for building. On the southern side of the plot, towards the water, the machine and boiler-houses, with their outer head structures and living quarters, formed the old ensemble of the power station.
On the main façade, towards the Spree, a clear reference was made to the formal style of late Gothic style, specifically, to the tradition of brickwork Gothic, characteristic of the Brandenburg March. Two powerful corner pylons with a portal-wall inserted between them embodied the

30-kV-Schalthaus, Fassadendetail, Aufnahme 1926
30 kV Switching house, façade detail, photograph 1926

Abbruch alte Kesselhausfassade 1954
Demolition of old boiler-house façade 1954

Die Halle zeigt noch weitgehend das Erscheinungsbild aus der großen Modernisierungsphase der 20er Jahre. Über einem Untergeschoss erstreckt sich der stützenfreie Raum auf eine Länge von 136 Metern. Überwölbt wird die Halle durch eine segmentbogenartige Stahlbinderkonstruktion mit Rautenoberlichtern. Die letzte Erweiterung von 1929 setzt sich als kubischer Anbau formal deutlich von der übrigen Halle ab. Der Wandaufbau an den beiden Längsseiten ist mit vorgeblendeten Bogenstellungen zur Aufnahme der Kranbahnschienen gegliedert.
Trotz der Veränderungen im Laufe der langen Entwicklungsgeschichte des Kraftwerks hat sich das Herzstück der alten Kraftwerksanlage, das Maschinenhaus mit angrenzendem Wohnhaus, in den wesentlichen räumlichen und visuellen Merkmalen erhalten. Das Ensemble darf mit seiner der Spätgotik entliehenen Fassadenarchitektur zu den herausragenden Zeugnissen der Industriearchitektur des Historismus gerechnet werden.

Das 30-kV-Schalthaus

Von den Erweiterungsbauten aus den 1920er Jahren sind die beiden Schalthäuser von besonderer architektonischer Bedeutung. Das 1926 nach einem Entwurf des Architekten Schönburg errichtete 30-kV-Schalthaus wird dominiert durch seinen expressiv ausgeformten Kopfbau zur Wasserseite. Die fünfgeschossige Klinkerfassade zur Spree ist geprägt durch einen überhöhten Mittelteil, dessen vertikale Fensterbänder durch vorspringende Mauerwerkspfeiler gerahmt werden, sowie durch spannungsreiche Ecklösungen, bei denen die Gebäudekanten in Fenster aufgelöst und gleichzeitig durch kräftig auskragende Betonvordächer betont werden. Der Bau wird durch ein Basisgeschoss mit glatten Klinkerflächen, großformatigen Fenstern und durch weit ausladende Dachgesimse gefasst. Ähnlich wie das 1928 entstandene 110-kV-Schalthaus an der Nordseite des Grundstückes befindet sich das Gebäude im Spannungsfeld zwischen Expressionismus und Neuem Bauen, weist jedoch eine mehr sachlich-moderne Formensprache auf.

Die Erweiterungsbauten der 1950er Jahre – Der Neubau des Kesselhauses

In den Jahren zwischen 1954 und 1966 wurde im Zuge der grundlegenden Modernisierung der technischen Anlagen das alte Kesselhaus mit seiner Prachtfassade zur Spree abgebrochen und in mehreren

great compositional theme and signaled the division of the ensemble into a trio that was in keeping with their hidden functions. The interplay, rich in contrasts, between red brick and lighter, plaster-work surfaces was successfully orchestrated here. Between the two tower projections reserved structure of the gabled hall. The central machine shop dominated the façades of the boiler-house and the living accommodation, with extended and towering themes.
The high, aesthetic standard expected of the exterior was to match the interior of the hall. The structural highpoint was the switching gallery on the end wall. Behind the middle three, pointed archways, decorated with striking Tudor gables, the central purpose of the switching unit was almost irreverently emphasised. The two exterior openings conveyed the transition to the head structure with the administrative wing. Nowadays, the arches of the gallery are closed. The hall still largely appears the same as in the great wave of modernisation in the 1920s. The free space stretches across a lower ground-floor level to a length of 136 metres. Above the hall is an archway that covers the space in a segmented, steel-truss structure with angular-shaped lighting. The last modification in 1920 a cubic extension, is clearly to be formally separated from the remainder of the building. The wall on each of the long sides is designed to take tracks for a crane, with arches set into the structure.
Despite the modifications in the course of its long history, the centrepiece of the old power station site, the machine shop and neighbouring accommodation has essentially retained its spatial and visual features. The ensemble, with its architectural main façades borrowed from late Gothic style, can be counted amongst the foremost monuments to the industrial architecture of a more traditional style.

The 30 kV Switching House

The two switching houses, emerging from the extensions of the 1920s, are of special architectural significance. The 30 kV switching house, built in 1926, to a design by the architect, Schönburg, dominates with its expressively formed head-structure to the water's edge. The five-storey clinker façade to the Spree is characterised by a raised middle section, with vertical window-edges, framed by protruding stonework pillars as well as emphasised by high-tension cornice solutions with

Kraftwerksensemble mit REA, Aufnahme Ende der 1990er Jahre
Power station ensemble with REA, photograph, end of 1990s

Baustufen durch einen Neubau ersetzt. Verantwortlich für den Entwurf war die Bewag-Bauabteilung unter der Leitung von Willy Rathge. Der Stahlskelettbau knüpft mit seiner roten Klinkerhülle zur Uferfront vereinfachend an die schlichte Formensprache der 20er Jahre an, zeigt sich aber von monumentalisierenden Tendenzen beeinflusst, die im Industriebau der 30er Jahre vorherrschten. Durch schmale Lisenen gegliederte vertikale Fensterbänder werden von einem großflächig-glatten Klinkermauerwerk gefasst und geben der Fassade eine strenge Geschlossenheit. Im Nebeneinander mit der historisierenden Fassade des Maschinenhauses hat der hoch aufragende, blockhafte Baukörper die Silhouette entscheidend verändert und markiert zusammen mit dem aus den 1920er Jahren erhaltenen 125 Meter hohen Schornstein einen städtebaulichen Akzent innerhalb des Kraftwerksensembles.

Die letzte Modernisierungsphase

Als letzte Erweiterung auf dem Kraftwerksgelände entstand 1989 der voluminöse, fensterlose Kubus der Rauchgasentschwefelungs- und Entstickungsanlage nach Plänen von Walter Henn. Auf der Südostecke der Fassade wird das Motiv des 30-kV-Schalthauses als Relief zitiert. Mit seiner Ausdehnung und Höhenentwicklung führt der Bau den Standort in eine neue städtebauliche Dimension.

Zustand und Nutzung im Jahr 2003

Nach dem Ende der konventionellen Stromerzeugung im Jahr 2001 dokumentiert das heutige Erscheinungsbild, insbesondere die das Stadtbild prägende Silhouette zur Spree, die hundertjährige Entwicklungs- und Baugeschichte des Kraftwerks. Der überwiegende Teil der Anlagen wird heute nicht mehr benötigt. Die Gebäudehüllen stehen leer, Restfunktionen sollen verlagert werden. Der technische Betrieb wird sich künftig auf etwa 40 Prozent der Grundstücksfläche im nördlichen Bereich zur Quedlinburger Straße konzentrieren. Disponibel wird das gesamte Ensemble an der Spreeseite mit Maschinenhaus, angrenzendem Wohnhaus einschließlich der Erweiterungen an der Ostseite des Grundstückes, dem Kesselhaus mit Schornstein und der Werkstatt in den Rudimenten des alten Kesselhauses und ehemaligem Wohlfahrtsgebäude, dem 30-kV-Schalthaus mit Werkstatt- und Lagergebäude sowie der Rauchgasentschwefelungs- und Entstickungsanlage.

inlaid windows at the edge of the building, at the same time as strongly exposed cement canopies. The building is characterised by a basic floor-level with smooth, clinker surfaces, large windows and generous roof ledges. Similar to the 110 kV switching house, built in 1928, on the north side of the site, the building conveys elements both of expressionism and "Neues Bauen" style, although it also displays a more objective and modern formal style.

The Extensions of the 1950s – the Reconstruction of the Boiler-House

From 1954 to 1966, the old boiler-house was replaced by a new building. In a wave of basic modernisation of technical plants, the old, damaged main façade towards the Spree was replaced in several construction phases. The Bewag construction department was responsible for the design, directed by Willy Rathge. The skeleton steel structure, with its red clinker shell, joins with the river frontage in a simple style of clear lines, as in forms from the 1920s. Nonetheless, the design is influenced by trends for monumental forms that were dominant in the industrial buildings of the 1930s. Vertical window-edges are captured in narrow-shaped anchor blocks, by a smooth, large surface area of clinker stonemasonry, giving the façade its strict closed form. Contrasting with the traditional looking façade of the machine shop, the block-style and tall building structure decidedly altered the silhouette and together with the 125 metre chimney – still standing from the 1920s – it defines an element of urban construction within the grouping of the power station.

The Last Modernisation Phase

The last extension of the power station site occurred in 1989, with the large-scale, windowless cube of the sulphur and nitrogen flue-fume removal plant. The plans were by Walter Henn. On the south-eastern corner of the façade, the motif of the 30 kV switching house is cited as a relief pattern. With its extension and height, the structure takes this location to a new dimension of urban construction.

Condition and Usage in 2003

In 2001, after the end of conventional electricity production, today's image, especially the characteristic silhouette defining the city towards

Kraftwerk Charlottenburg • Charlottenburg Power Station, Am Spreebord 5-7, 10589 Berlin					
Grundstücksfläche gesamt		Land Area, in total			69.589 m²
Vermarktbare Flächen		Marketable Sites			
Grundstücksfläche		Land Area			46.000 m²
Bruttogeschossfläche (BGF) gesamt		Gross Floor Area, in total (BGF)			73.000 m²
darunter:	Denkmalensemble	including:	Listed Building Ensemble		33.000 m²
mit:	Maschinenhalle	with:	Machine Hall		10.600 m²
	Kesselhaus		Boiler-House		15.000 m²

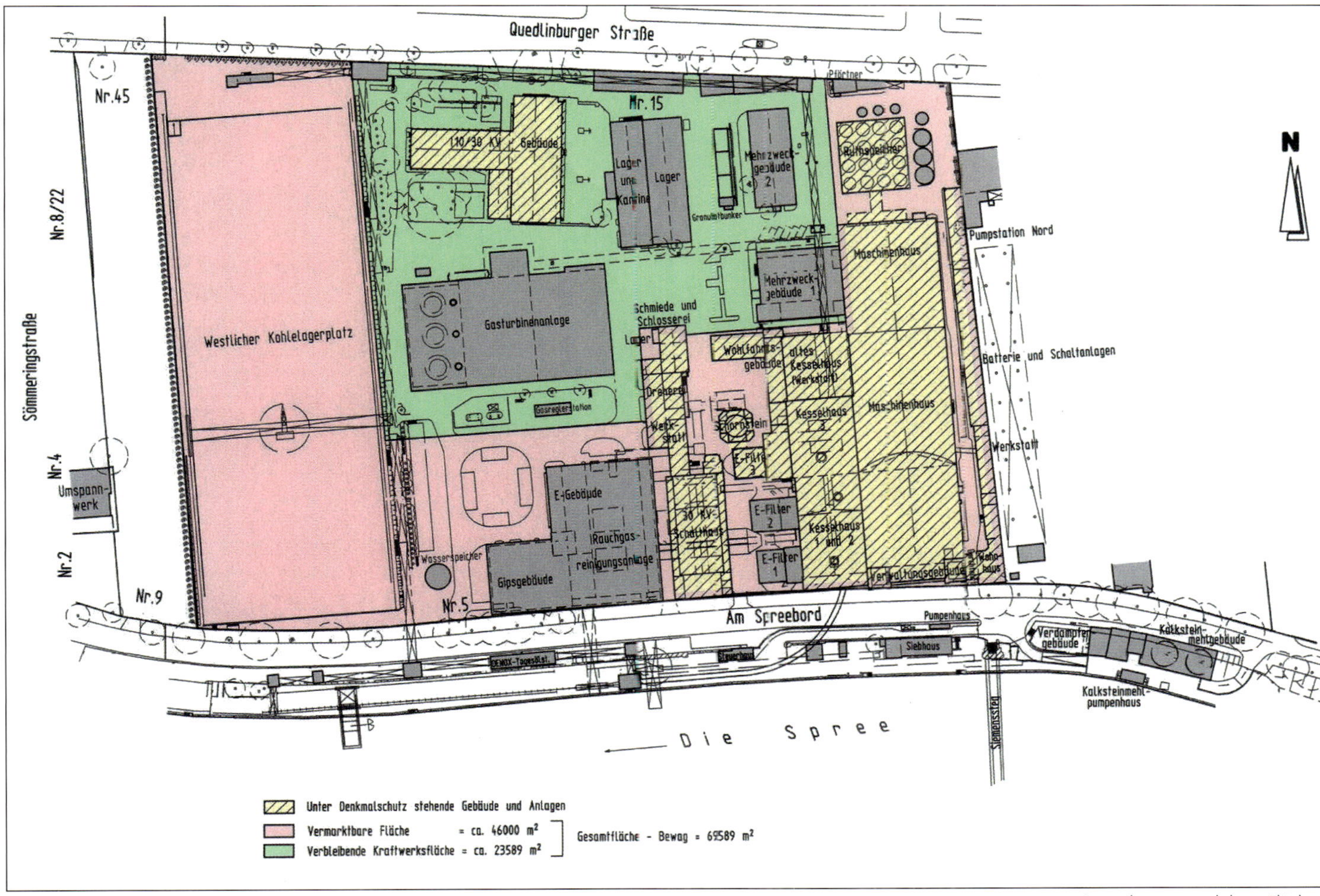

Lageplan, vermarktbare Flächen
Site plan, marketable land

Der Kern der alten Kraftwerksanlage, das Maschinenhaus mit angrenzendem Wohnhaus, hat sich, wenn auch teilweise verändert, in seinen architektonischen Merkmalen bis heute erhalten. Die historisierende, neugotische Fassade der Halle wurde im Laufe der Zeit vereinfacht, die beiden Turmhelme und die eine Fackel tragende Statue auf der Spitze des Giebels sind verloren. Im Innenraum ist die technische Ausstattung aus der zweiten Modernisierungsphase der 60er Jahre vorhanden, im Nordanbau stehen noch die beiden Speicherturbinen für die Ruths-Speicher von 1929.

Das 30-kV-Schalthaus von 1926 weist überwiegend den ursprünglichen technischen Ausbauzustand auf. Der Geschossbau ist in Grundrissdisposition und äußerem Erscheinungsbild weitgehend erhalten. Der Kesselhausneubau ist im Erscheinungsbild seiner Entstehung erhalten, sein Innenraum von 41 und 51 Metern Höhe wird dominiert von den Kesselanlagen und -ebenen der 60er Jahre. Der westliche dominante Eckpunkt der Kraftwerkssilhouette, die 1989 errichtete Rauchgasentschwefelungsanlage, steht außen und innen unverändert da. Da die Anlage nicht unter Schutz steht, wird ihr Erhalt unter Betrachtung der städtebaulichen und architektonischen Qualitäten sowie der ökonomischen und ökologischen Ressourcen abzuwägen sein.

the Spree, provides a record of a century's development and the construction history of the power station. The majority of the sites are no longer needed today. The shells of buildings stand empty; and remaining operations is to be transferred elsewhere. In future, technical operation will be concentrated on about 40 per cent of the land in the northern area of the site towards Quedlinburger Straße. The entire ensemble that is available for use towards the Spree includes the machine shop, the adjacent accommodation, including extensions on the site's eastern side, the boiler-house with chimney and the workshop with foundations of the old boiler-house and the former welfare building, the 30 kV switching house with workshop and storage building, as well as the sulphur and nitrogen flue-fume removal plant.

The core of the old power station site, the machine shop with adjacent accommodation, has only partly been modified and to this day, it retains its specific architectural features. Over time, the traditional-looking, neo-Gothic façade of the hall structure was simplified, the two tower-helms and the torch-bearing statue on the top of the gable were lost. In the interior of the building, technical equipment from the second modernisation phase, in the 1960s, is preserved; and in the north

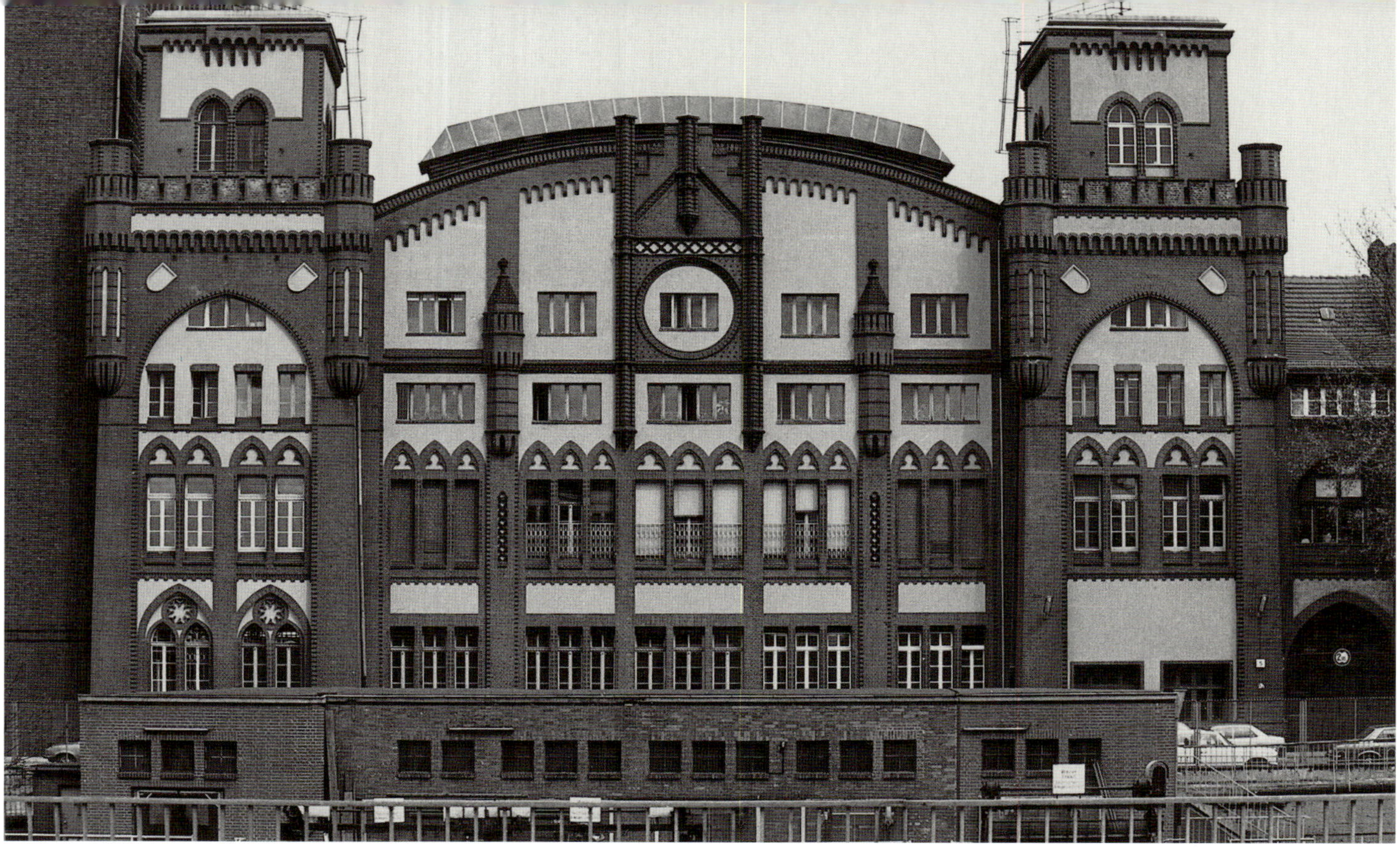

Ansicht Maschinenhalle, Kopfbau, aktuelle Aufnahme
View of machine hall, frontage, up-to-date photograph

Die planungsrechtliche Situation

In der Neufassung des Flächennutzungsplanes (FNP) aus dem Jahr 2002 sind der Kraftwerksstandort Charlottenburg und die westlich angrenzenden Flächen bis zur Sömmeringstraße als Ver- und Entsorgungsfläche mit gewerblichem Charakter, Zweckbestimmung Energie, dargestellt, die Flächen östlich angrenzend bis zur Darwinstraße als gemischte Baufläche. Entlang des Spreeufers ist ein durchgängiger Grünzug vorgesehen.

Bei der geplanten Reduzierung der betriebsnotwendigen Flächen auf einen Kernbereich des heutigen Kraftwerksgeländes ist von einer FNP-Änderung auszugehen.

Planungsrechtlich ist der Baunutzungsplan von 1960 maßgeblich, hier ist der Kraftwerksstandort als reines Arbeitsgebiet mit Baustufe V/3 ausgewiesen. Für die östlich angrenzenden Flächen und für das westlich der Sömmeringstraße sich erstreckende Gelände liegen rechtskräftige Bebauungspläne vor.

Die Festlegungen der verbindlichen Bauleitplanung werden mit der angestrebten Entwicklung des Gebietes zu modifizieren sein.

extension both of the accumulator turbines for Ruth's 1929 accumulator-unit are still standing.

The 1926 30 kV switching house also displays major elements of its original technical status. The outline of the ground-floor level is largely preserved according to the original plan and exterior appearance. The newly constructed boiler-house is preserved as created, with its interior space, to a height of 41 and 51 metres, dominated by boiler installations and levels from the 1960s. The dominant point on the western corner of the power station's silhouette, the 1989 sulphur flue-fume removal plant, is unchanged in both external and internal appearance. Since this site is not listed land, its preservation is to be evaluated by paying due regard to urban construction and architectural aspects, as well as economic and ecological resources.

The Legal Planning Situation

In the new version of the land utilisation plan (FNP), dating from 2002, the site of the Charlottenburg Power Station and the adjacent plots to the west as far as Sömmeringstraße, are described as supply and waste removal land, with commercial value for energy purposes, and the adjacent plots to the east, as far as Darwinstraße, are noted as a mixture of building land. A continuous green belt is planned along the banks of the river Spree.

The planned reduction of commercially designated areas on a core sector of today's power station site is likely to include an FNP modification.

In legal planning terms, the 1960 utilisation plan continues to set the standard, according to which the location of the power station is designated purely as a commercial area, with building level V/3. For the adjacent sites to the east and for the site extending on the western Sömmeringstraße, building plans are available that are legally binding. The legal provisions for site management are to be modified according to the desired development of the area.

Maschinenhalle, Aufnahme 1991
Machine hall, photograph 1991

Kraftwerksensemble an der Spree, Aufnahme 1954
Power station ensemble at the river Spree, photograph 1954

Nutzungs- und Entwicklungspotenziale

Mit der Einstellung des konventionellen Kraftwerksbetriebes steht der Standort Charlottenburg an der Schwelle zu einem radikalen Strukturwandel und wird damit als Motor für die Entwicklung der benachbarten brachgefallenen Flächen wirken.
Die Grundlage für die Entwicklung und Öffnung des Gebietes wird der überarbeitete Bereichsentwicklungsplan (BEP) sein. Darüber hinaus hat die Bewag in Zusammenarbeit mit dem Bezirk und in Vorbereitung eines Bebauungsplanverfahrens eine Entwicklungskonzeption „Am Spreebord" erstellt, die Entwicklungsziele definiert und Möglichkeiten zur Aufwertung des Standortes und der Umgebung aufzeigt. Der „Spreebord" soll durch die Anlage eines Ufergrünzuges als Lückenschluss zwischen Innenstadt/Tiergarten und Schlossgarten als öffentlicher Raum qualifiziert werden.
Für das Kraftwerksgelände eröffnet sich ein breites Spektrum von Zukunftsoptionen, die von Gewerbe-, Dienstleistungs- und Einzelhandelsnutzungen bis zu den Bereichen Kultur/Bildung/Sport, Freizeit/Begegnung sowie Gastronomie- und Hotelwesen reichen. Die Ergebnisse eines studentischen Ideenwettbewerbs zur Entwicklung denkmalgerechter und wirtschaftlich tragfähiger Konversionskonzepte für die aus ihrer Nutzung gefallenen Kraftwerksflächen belegen dies eindrucksvoll.
Der Wettbewerb war gemeinsam von der Bewag und der Deutschen Stiftung Denkmalschutz unter der Schirmherrschaft des Berliner Landesdenkmalamtes im Herbst 2002 ausgelobt und im April 2003 mit großer Resonanz abgeschlossen worden (Dokumentation und Katalog, Herbst 2003). Die Arbeiten können Basis und Ansatz für konkrete Nachnutzungen und Investorengespräche liefern.
Das Kraftwerksensemble bestimmt mit seinen das Stadtbild prägenden Industriebauten den Genius Loci des Ortes und wird dadurch wesentlich seine künftige Entwicklung tragen.

Utilisation and Development Potential

With the introduction of conventional power station operations, the Charlottenburg location is on the threshold of radical structural change and, therefore, is to serve as a catalyst to develop the adjacent sites that lie obsolete.
The basis for the development and access to the area are to be the revised area-development plan (BEP). Apart from that, in cooperation with the district authority, the Bewag, has put together a development concept, "Am Spreebord", as a blue-print for the process of planning new building. This planning concept defines development goals and highlights the possibilities of improving the location and the surrounding area. The "Spreebord" project is meant to qualify as a public space, with its green-belt riverbank plot filling the gap between Berlin's inner-city area with the Tiergarten and the castle garden.
A broad spectrum of future options arises for the power station site. The options range from commercial, service and retail purposes to areas of culture, education, sport, leisure and meeting venues as well as hotel and catering. The results of a student competition for ideas on conversion projects that are both viable economically and respect the listed sites are an impressive indication of how to use obsolete power station sites.
In autumn 2002, the competition was co-sponsored by the Bewag with the German Foundation for Monument Preservation (Deutsche Stiftung Denkmalschutz), under the patronage of Berlin's State Monuments Authority. In April 2003, the competition was closed after a magnificant response (documentation and catalogue, autumn 2003). The projects may provide a basis and starting point for concrete re-utilisation schemes and for negotiations with investors.
The power station ensemble, with its industrial buildings that shape the city image, defines the spirit of the location and, for that reason, it will essentially define the future development of this site.

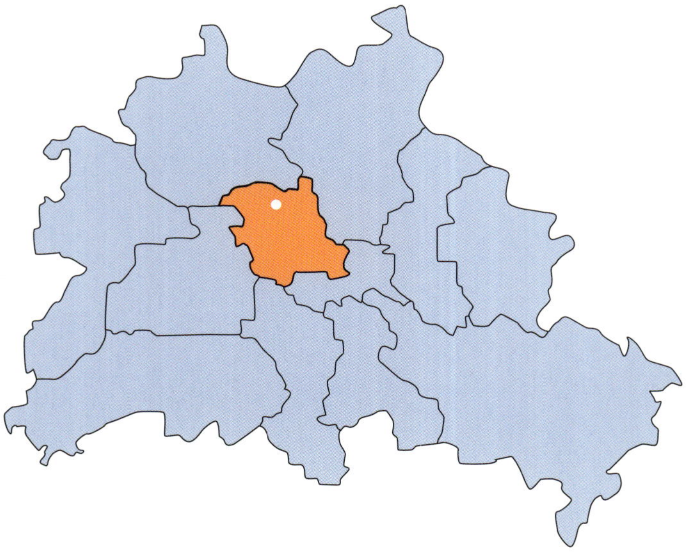

Kraftwerk Moabit
Moabit Power Station

DAS KRAFTWERK MOABIT
MOABIT POWER STATION
Friedrich-Krause-Ufer 10-15

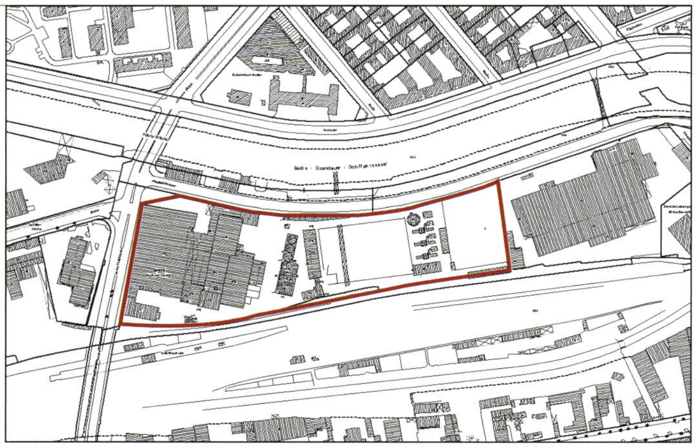

Lageplan
Site plan

Die stadträumliche Einbindung

Das Kraftwerk Moabit nimmt eine stadträumliche Schlüsselstellung als prägender Bestandteil der ausgedehnten Industrielandschaft zwischen Westhafen und dem Berlin Hauptbahnhof - Lehrter Bahnhof ein und wird gleichzeitig als markanter Orientierungspunkt in der heterogenen Stadtlandschaft im Übergang zwischen Wedding und Moabit erlebt.
Mit seiner Lage an der Putlitzbrücke zwischen den Gleisen der Ringbahn und dem Berlin-Spandauer-Schifffahrtskanal dominiert die Silhouette der Kraftwerksanlage den Stadtraum weithin.
Beidseitig des Industriegürtels erstrecken sich die dichten Bebauungsstrukturen von Moabit und Wedding. Nordwestlich des Kraftwerks liegt das Areal des Universitätsklinikums Rudolf Virchow, das zum Volkspark Rehberge mit der Gedenkstätte Plötzensee überleitet.
Der Standort ist durch den in der Nähe gelegenen Autobahnanschluss mit Anbindung an den Flughafen Tegel sowie durch ein dichtes Netz des Nahverkehrs mit unmittelbaren U- und S-Bahnanschlüssen verkehrstechnisch sehr gut erschlossen.

Die Entwicklung des Kraftwerks

Vorgeschichte

Das Jahr 1899 markierte in der Entwicklung der Berliner Elektrizitätsindustrie einen Entwicklungssprung, nachdem sich im letzten Jahrzehnt des 19. Jahrhunderts die Nachfrage nach elektrischer Energie in der aufstrebenden Metropole exorbitant entwickelt hatte. In dieses Jahr fiel ein neuer Vertrag zwischen dem Magistrat und den Berliner Elektrizitätswerken (BEW). Er sicherte der Gesellschaft das Wegerecht für ganz Berlin und hob die bis dahin bestehende Begrenzung der Kraftwerksstandorte auf das Stadtgebiet auf.
Die Bedeutung dieses Vertragsabschlusses, der es den BEW erlaubte, in Verbindung mit der neu entwickelten Drehstromtechnologie den Schwerpunkt der Stromerzeugung von der Innenstadt auf den Stadtrand zu verlegen, zeigte sich in den unmittelbar darauf einsetzenden

Urban Integration

The Moabit Power Station has a key urban status as a defining part of the extended industrial landscape between Westhafen, Berlin main station and Lehrter station. At the same time, it is seen as a defining geographical landmark on the diverse urban landscape crossing from Wedding to Moabit.
Due to its location on Putlitz bridge, between the railway and the Berlin-Spandau shipping canal, the power-station's site and silhouette overwhelmingly dominates the urban area.
The densely built structures of Moabit and Wedding stretch out on both sides of the industrial belt. To the north-west of the power station is the area of the Rudolf Virchow University Clinic that leads to the Rehberge Volkspark with the Plötzensee Memorial.
The site is easily accessible by an integrated nearby motorway network, with connection to the airport, Berlin-Tegel, and a well-integrated local transport network, with nearby underground and S-bahn, or city train connections.

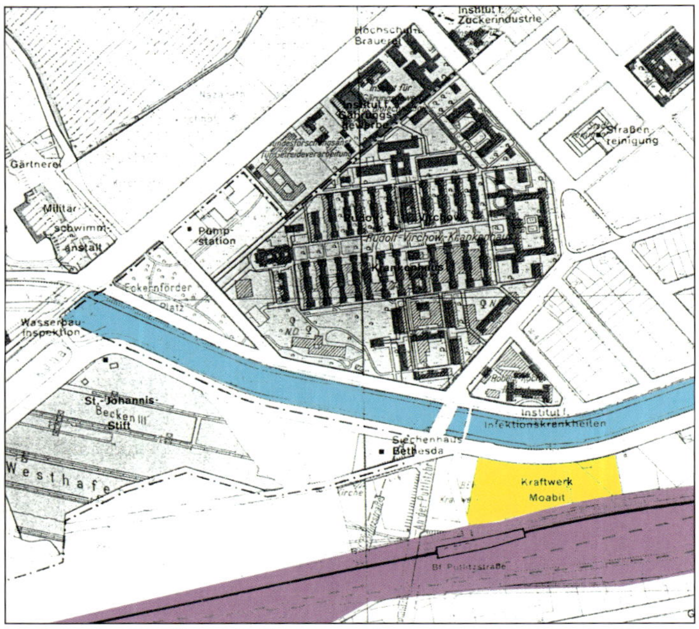

Übersichtsplan ca. 1910 1:15000
Overview plan around 1910

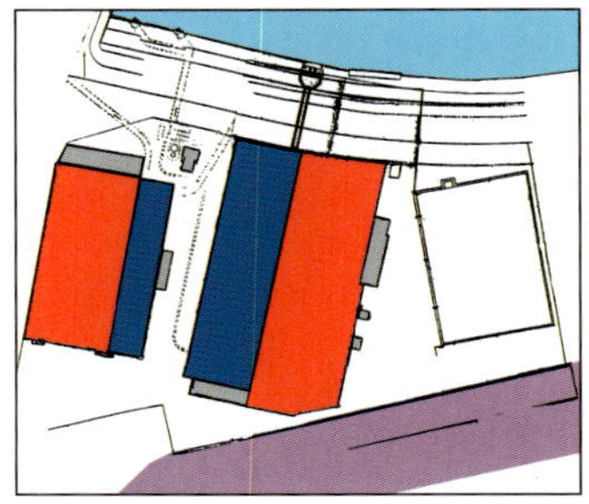

1913

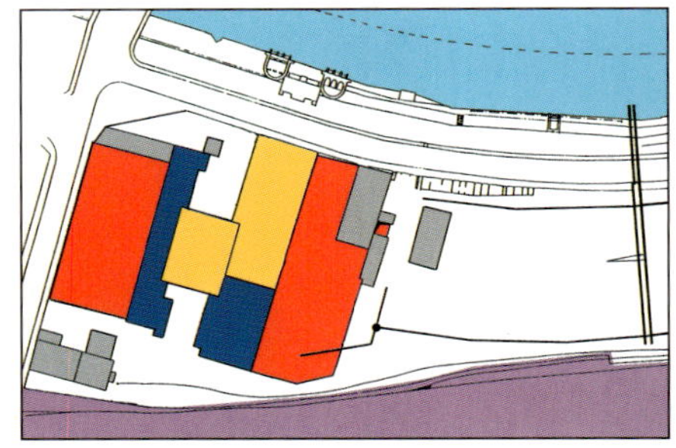

1925

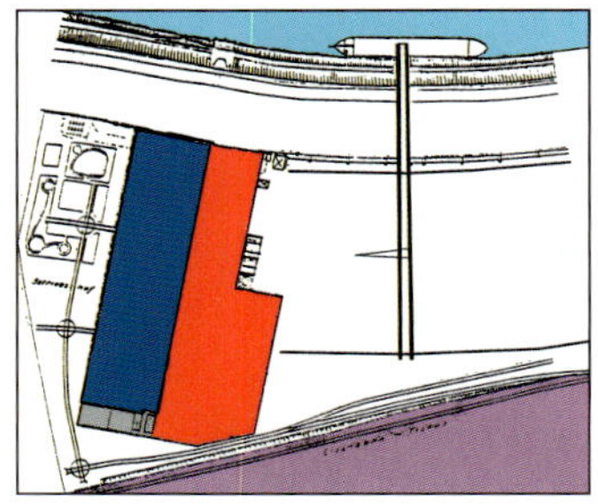

1900

Entwicklungsstufen des Kraftwerks
Power station's development in phases

Baumaßnahmen. Noch im selben Jahr begann die Planung für Moabit, 1900 erfolgte die Betriebseröffnung des Kraftwerks, das ebenso wie das fast zeitgleich entstandene Gemeindekraftwerk Charlottenburg als Drehstromkraftwerk errichtet wurde.
Mit einer Anfangsleistung von 7.200 kW belieferte das Kraftwerk Moabit das nördliche Stadtgebiet und speiste die Außenbereiche Reinickendorf, Pankow, Plötzensee und Spandau. Es war Ergänzung und Pendant zum Kraftwerk Oberspree, das als erstes Drehstromkraftwerk seit 1897 den östlichen Stadtraum versorgte und dessen Netz sich seit 1899 nach Übernahme durch die BEW in die südöstlichen Außenbereiche erweiterte.
Der gewählte Standort am Nordrand des industriereichen und dicht bevölkerten Ortsteils Moabit zwischen dem Berlin-Spandauer-Schifffahrtskanal und den Gleisen der Hamburg-Lehrter-Eisenbahn gewährte eine optimale Ver- und Entsorgung des Kraftwerks über Wasser und Schiene und bot die entsprechende Fläche für künftige Erweiterungen. Das Grundstück war in einer Gewerbezone angesiedelt, die als traditioneller Umschlagplatz zwischen den Bezirken Tiergarten, Wedding und Berlin-Mitte lag. Damit war sowohl die Nähe zu den Industrieballungen in Moabit und Wedding als auch zu den städtischen und privaten Abnehmern der expandierenden Metropole gesichert.

Moabit I von 1900 bis 1907

Mit dem Bau des neuen Drehstromkraftwerks, der Zentrale Moabit, wurde 1899 begonnen. Dass der renommierte Architekt Franz Schwechten mit der architektonischen Gestaltung des Kraftwerks Moabit beauftragt wurde, spiegelte das wachsende Selbstbewusstsein der jungen Elektrowirtschaft, das auch in einer anspruchsvollen Architektursprache zur Geltung gebracht werden sollte.
Schwechten zählte um 1900 zu den erfolgreichsten Architekten der Reichshauptstadt und hatte in den 1890er Jahren die Position als gestaltender Architekt bei der AEG inne.
Der erste Bauabschnitt entstand unmittelbar am Friedrich-Krause-Ufer. In seiner räumlich-funktionalen Disposition folgte das Kraftwerk Moabit der bewährten Parallel-Aufstellung von Kessel- und Maschinenhaus, die als entwicklungs- und erweiterungsfähig galt. Von diesem Konzept wurde erstmals 1910/11 beim Bau des Kraftwerks Steglitz abgewichen. Ausgerüstet wurde das Kraftwerk mit modernsten,

The Development of the Power Station

Pre-History

The year 1899 marked a leap ahead in the development of Berlin's electricity industry, after an massive increase in demand for electricity in this up-and-coming metropolis in the final decade of the 19th century. This was the year that a new contract was signed between the Magistrate's Office and Berlin's Electricity Works (BEW). It secured the BEW rights for the whole of Berlin and suspended the previously existing restriction of power station sites in urban areas.
The significance of this contract, allowing the BEW, in the context of the newly developed alternating current technology, to transfer major electricity production from the city centre to the periphery, is evident in the building measures that were taken immediately after the contract was signed. In the very same year, plans were made for Moabit that opened in 1900. This power station was constructed as an alternating current plant, just as the power station in the parish of Charlottenburg, which was constructed at almost the same time.
Moabit Power Station had a starting capacity of 7,200 kW and supplied the northern part of the city, in particular the outer limits of Reinickendorf, Pankow, Plötzensee and Spandau. This plant was both a complement and contrast to Oberspree Power Station, the first alternating current works that supplied the eastern sector of the city from 1897, with its network being expanded from 1899, after the BEW takeover of the south-eastern, outer city limits.
The chosen location on the northern edge of the industrial and densely populated district of Moabit, between the Berlin-Spandau shipping canal and the Hamburg-Lehrter railway provided optimal supply and waste removal routes to the power station, by water and rail, as well as appropriate sites for future expansion. The land was situated in a commercial area, located in a traditional transfer route between the city districts of Berlin Tiergarten, Wedding and Berlin-Mitte. In that way, proximity both to Moabit's and Wedding's industrial centres and to urban and private suppliers was secured for the expanding metropolis.

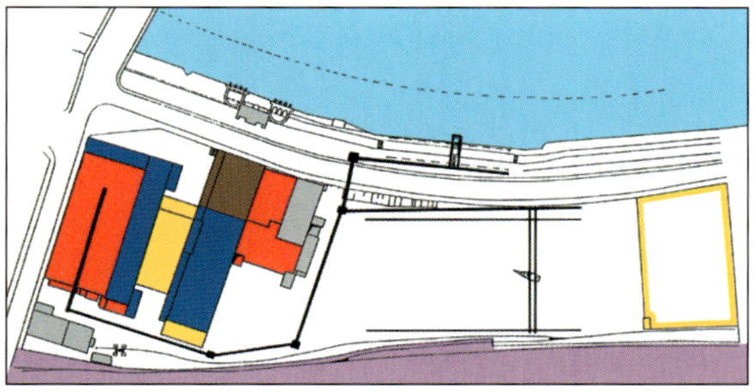

1953

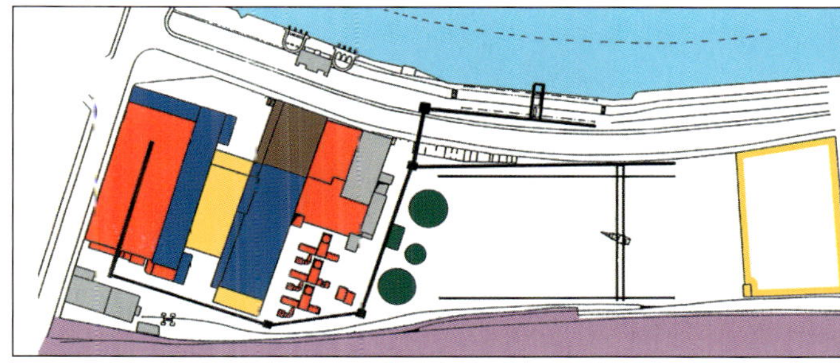

1972

2000

●	Wasserstraßen	Waterways
●	Bahngelände	Railway Area
●	Kesselhaus	Boiler-house
●	Maschinenhaus	Machine Hall
●	Stromverteilung	Electricity Distribution
—	Freiluftanlagen	Open Air Sites
●	Heizölversorgung	Heating Oil Supply
●	Umweltschutz	Environmental Protection
●	Nebengebäude aller Art	Various Outbuildings
●	Kohlenversorgung	Coal Supply

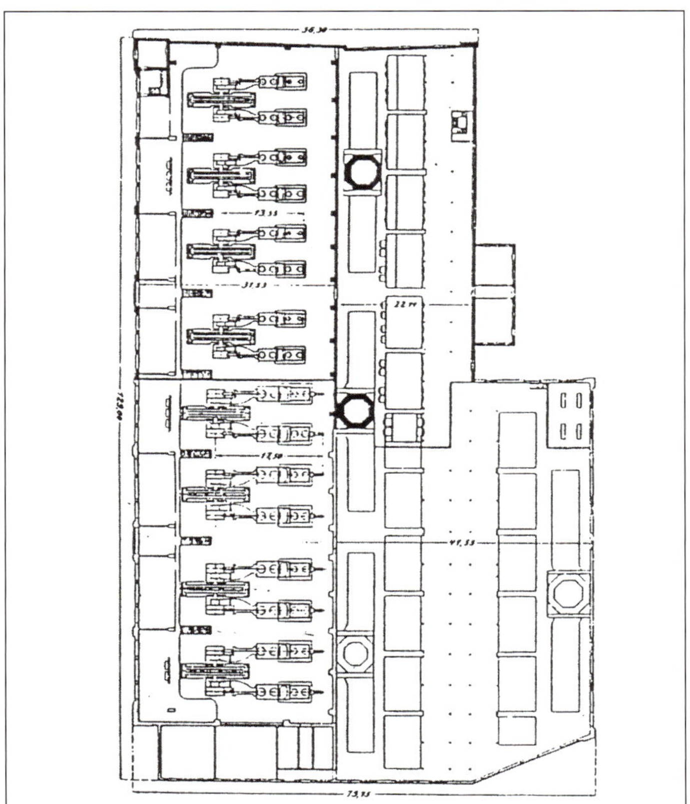

Moabit I, historischer Grundriss
Moabit I, historical ground plan

stehenden Sulzerschen Dampfmaschinen. Noch vor der Inbetriebnahme des ersten Bauabschnittes musste die Anlage vergrößert werden; die Ausrüstung wurde zunächst auf acht Dampfmaschinen erweitert. Die großtechnische Erprobung der Dampfmaschine leitete die Rolle des Kraftwerks Moabit als Experimentierfeld der BEW für technische Neuerungen ein.

1906 nahm ein erster Turbosatz von drei Dynamos den Betrieb auf. Mit der Aufstellung der Turbinen fand die Entwicklung des Kraftwerks nach Süden ihr bauliches Ende, die maximale Ausbaukapazität war erreicht.

Moabit I from 1900 to 1907

Construction on the new alternating current power station, Central Moabit, commenced in 1899. The famous architect, Franz Schwechten, was commissioned to design of Moabit Power Station, and this was a reflection of the increasing self-confidence of the fledgling electrical industry that was also to be supported in sophisticated, architectural expression.

Around 1900, Schwechten was amongst the capital's most successful architects and in the 1890s, he occupied the position of design architect at AEG.

The first construction phase was developed directly on Friedrich Krause riverbank. In terms of its spatial and functional design, Moabit Power Station was planned on the proven parallel boiler-house and machine shop model that was considered valid for development and expansion. Only in 1910/11 was this model set aside for the building of Steglitz Power Station. The installation was equipped with the most modern standing Sulzer steam machines. The site had to be expanded, even before the first construction phase was operational; and the machinery was first extended by a further eight steam machines. The large-scale technical testing of the steam machine saw Moabit's role as an experimental field for BEW's technical innovations.

In 1906, a first turbo-set of three dynamos went into operation. The installation of turbines put an end to the site's building development to the south and the power station reached maximum capacity.

Moabit I, Turbinen, Aufstellung 1906
Moabit I, turbines, installation 1906

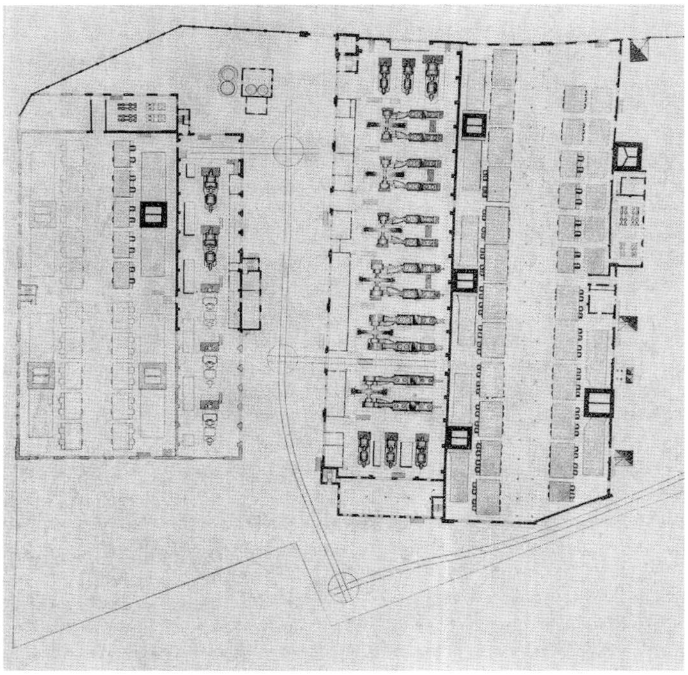

Zentrale Moabit I und II, historischer Grundriss
Central Moabit I and II, historical ground plan

30-kV-Schalthaus im Bau, Südseite 1925
30 kV Switching house, under construction, south side 1925

Moabit II 1908

Im Jahr 1908 begann mit dem Bau eines neuen Kraftwerks auf dem westlichen neu erworbenen Teil des Grundstückes neben der Putlitzbrücke ein neuer Abschnitt in der Geschichte der Zentrale Moabit. Entsprechend der technischen Entwicklung sollte Moabit II ausschließlich mit Turbinen zur Energieerzeugung ausgestattet werden. Ein Vergleich der Grundrisse von Moabit I und II macht die enorme räumliche Einsparung durch die Verwendung von Turbosätzen deutlich. Beide Kraftwerke waren in ihrer Leistungskapazität identisch, in ihrer Grundfläche unterschieden sich die nur acht Jahre auseinander liegenden Werke jedoch beträchtlich: Das neue Werk benötigte nur die Hälfte des Raumes. Der konstruktive Aufbau und die parallele Anordnung von Kessel- und Maschinenhaus folgten dem Vorbild von Moabit I. Der architektonische Entwurf lag wieder in den Händen von Franz Schwechten.

Auch Moabit II sollte in mehreren Abschnitten ausgebaut werden, der erste Bauabschnitt ging noch 1908 in Betrieb. Zu der Zeit galt das Kraftwerk Moabit als das größte und modernste Drehstromwerk Berlins.

Die Erweiterungen der 1920er Jahre

In den Kriegsjahren zwischen 1915 bis 1918 wurde das Werk Moabit II unter Berücksichtigung der ursprünglichen Planung erweitert. Nach dem Krieg verlor das inzwischen technisch veraltete Kraftwerk Moabit I schrittweise seine ursprüngliche Funktion und wurde zum „Abladeplatz" für neue technische Nutzungen, was den Gesamteindruck des Halleninneren beeinträchtigte. Die Westfassade erfuhr durch eine Aufstockung einen wesentlichen Eingriff.

Um 1925 wurden die Werke Moabit I und II durch einen fünfgeschossigen Riegelbau verbunden, in dem eine 30-kV-Schaltanlage untergebracht war. Dieser Bau wurde durch die Umstellung der Verteilungsspannung von sechs auf 30 kV notwendig und durch das Baubüro der Bewag konzipiert.

Mitte der 1920er Jahre erfolgte mit der Umstellung des Kraftwerksbetriebes auf Kohlenstaubfeuerung die letzte Erweiterung von Kesselhaus II nach Süden. Das Werk nahm erst jetzt die 1908 geplante Grundfläche ein. Auf der Südseite von Moabit II entstand 1925 die vom damaligen Bewag-Chefarchitekten Hans Müller entworfene Kohlen-

Moabit II 1908

In 1908, the construction of a new power station on the newly acquired western plot, led to a new phase in the history of Central Moabit. Located near Putlitz bridge and keeping pace with technical progress, Moabit II was to be only equipped with turbines for energy production. A comparison of plans for Moabit I and II shows how the use of turbine sets saved an enormous amount of space. Each power station had an identical output capacity, yet their surface area differed considerably, even though only eight years separated their construction: the newer power station needed half the space. The construction and parallel building of boiler-house and machine shops followed the example of Moabit I. The architectural design was again in the hands of Franz Schwechten.

Moabit II was also to be developed in several phases, the first phase became operational in 1908. At that time, Moabit Power Station was regarded as Berlin's largest and most modern alternating current plant.

The 1920s Extensions

During the war years, from 1915 to 1918, Moabit II was extended with respect to the original plans. After the war, Moabit I, by now technically outdated, gradually lost its initial purpose to become a "dumping ground" for technical innovations. The overall impression of the plant's exterior was adversely affected. The addition of another storey to the western façade was a major innovation.

Around 1925, the Moabit I and II works were connected by a five-storey half-timber construction that housed a 30 kV switching installation. This structure was designed by the Bewag's construction office, due to the conversion of the distribution network from six to 30 kV.

In the mid-1920s, the conversion of the power station's output to coal-dust firing led to the last extension of boiler-house II towards the south. Only now did the plant take up the planned 1908 surface area. On the southern side of Moabit II, the coal-dust grinding and processing unit was built to the 1925 design of former Bewag chief architect, Hans Müller. The introduction of coal-dust firing made Moabit Power Station a model for this new technical heating process. The project developed with such promise that the Bewag decided to equip the new high-capac-

Kohlenstaubmahl- und Aufbereitungsanlage im Bau, Aufnahme 1925
Coal-dust grinding and processing unit, under construction, photograph 1925

staubmahl- und Aufbereitungsanlage. Mit der Einführung der Kohlenstaubfeuerung wurde das Kraftwerk Moabit zum Modell für dieses neue heiztechnische Verfahren. Es entwickelte sich so Erfolg versprechend, dass sich die Bewag entschloss, das neue Großkraftwerk Klingenberg mit Großkesseln und Kohlenstaubfeuerung auszurüsten.
In den Jahren 1929/30 wurde durch die Architekten Waltar Klingenberg und Werner Issel das Werk Moabit II durch ein Umformerwerk mit Front zur Putlitzbrücke ergänzt und der alte Gebäudekomplex dreiseitig mit einheitlicher Fassadenfront überformt. Damit waren die wesentlichen baulichen Maßnahmen bis zum Zweiten Weltkrieg abgeschlossen.

Die Entwicklung nach dem Zweiten Weltkrieg
1943 wurde das Maschinenhaus II schwer beschädigt, weitere Kriegsschäden folgten. Nach den dringendsten Reparaturen der Nachkriegszeit begann Anfang der 50er Jahre eine grundlegende Modernisierung der technischen Anlagen, die insbesondere im Inneren der Hallen zu weit reichenden Eingriffen führte.

ity Klingenberg Power Station with high capacity boilers and coal-dust firing.
In 1929/30, Moabit II was expanded by a converter-unit, designed by architects, Waltar Klingenberg and Werner Issel. On three sides, the old building complex was re-modeled with a unified façade and building frontage. These measures concluded major construction until World War Two.

The Development after World War Two
In 1943, the machine shop was badly damaged and further war damage followed. After the emergency repairs during the post-war era, in the early 1950s, an overall modernisation of technical installations commenced, leading to far-reaching innovations, particularly in the halls' interior.
From 1952 to 1958, the power station was converted into a peak demand plant to replace the outdated plants of Moabit I and II. In unit II, new

Historische Turbine in der Ausstellungshalle, um 1990
Historical turbine in the exhibition hall, around 1990

Kohlenstaubkessel, 1926
Coal-dust boiler, 1926

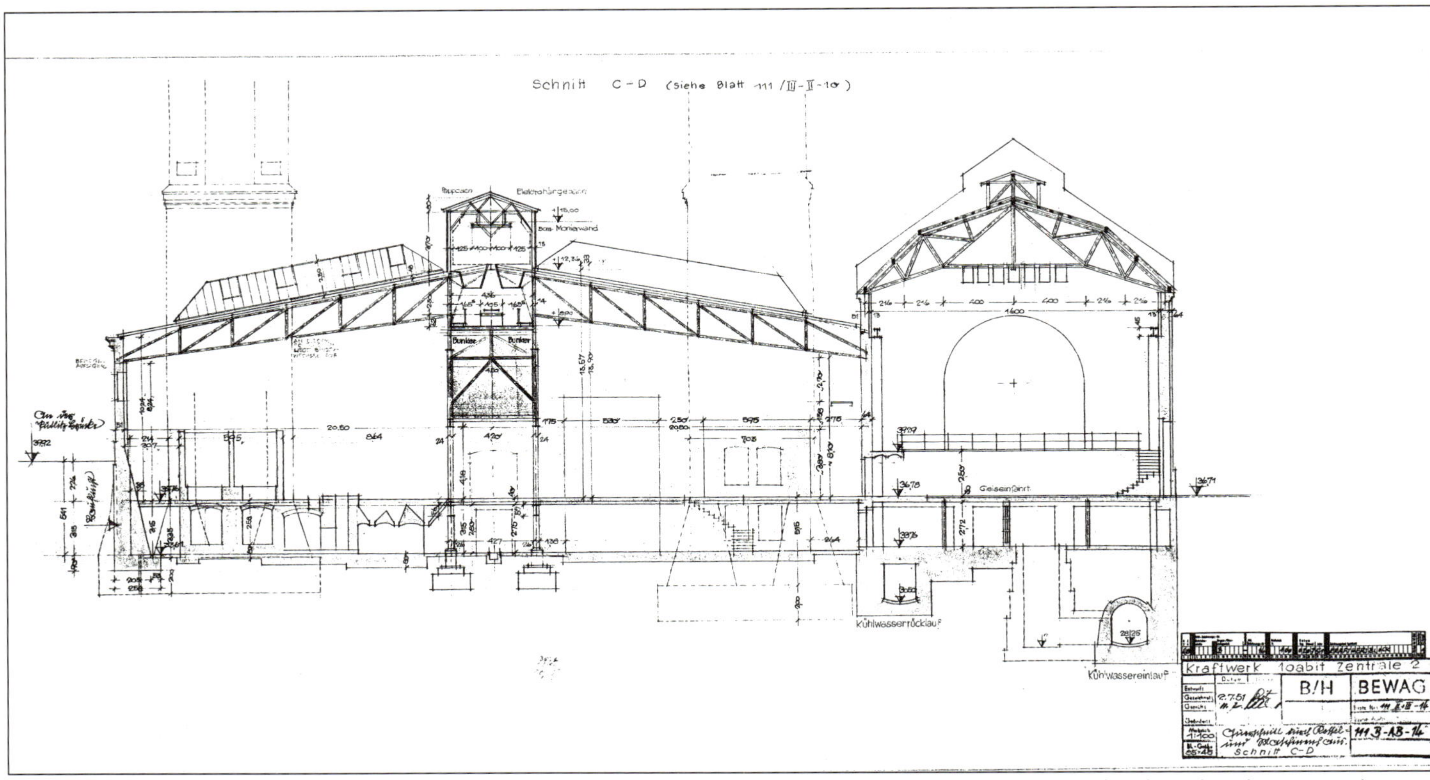

Moabit II, Querschnitt, Kessel- und Maschinenhaus, Zustand um 1957
Moabit II, cross-section, boiler-house and machine hall, condition around 1957

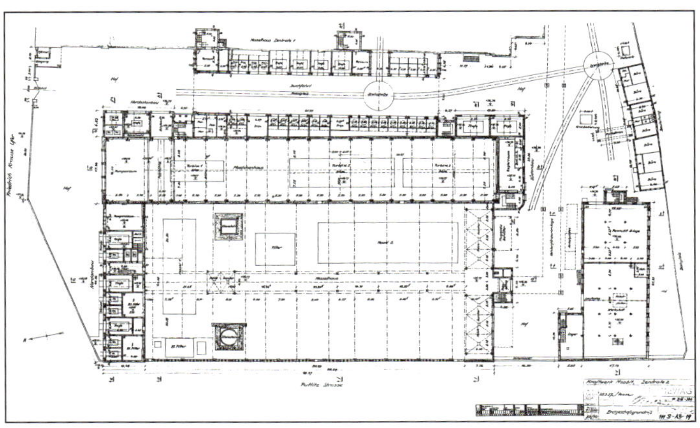

Moabit II, Grundriss, Zustand um 1957
Moabit II, ground plan, condition around 1957

Zwischen 1952 und 1958 wurde das Kraftwerk zur Spitzenlastanlage als Ersatz für die veralteten Anlagen von Moabit I und II ausgebaut. Drei neue Heizblöcke wurden im Werk II untergebracht. Um mit der vorhandenen Raumhöhe des Kesselhauses auszukommen, wurden spezielle liegende Kessel entwickelt. Zusätzlich wurde eine Dampfturbine von 1924 für Notfälle als Block IV an ihrem alten Standort im Werk I wieder in Betrieb genommen. Diese historische Turbine ist als Ausstellungsstück in Moabit erhalten geblieben. Mit einer Kapazität von 155.000 kW war die Modernisierungsphase 1958 abgeschlossen, die Zentrale Moabit offiziell als Heizkraftwerk in Betrieb gegangen.
Von besonderem Interesse unter den zahlreichen technischen Anlagen, die vor dem letzten großen Ausbau des Kraftwerks in den 1980er Jahren auf dem Gelände entstanden, ist die 1971 in Betrieb genommene Gasturbinenanlage am Standort des Kesselhauses I, dessen Südteil dafür abgerissen werden musste. Drei verkapselte Turbinen wurden frei ohne Verkleidung aufgestellt: die technische Gestalt wurde selbst zur Architektur.

heating blocks were housed. Special horizontal boilers were developed in order take advantage of height. In addition, a 1924 steam turbine for emergencies was located on its former site in plant I and was put into service again as Block IV. This historic turbine is preserved as a showpiece in Moabit. With a capacity of 155,000 kW, the 1958 phase of modernisation was complete. Central Moabit officially went into service as a thermal energy plant.
Amongst the numerous technical installations that emerged with the last substantial building on the site in the 1980s, the gas turbine unit is of particular interest. This unit dates from 1971 and went into service on the site of boiler-house I. The southern part was demolished for that purpose. Three capped turbines were built, without casing: the technical form was itself architecture.

Moabit I, Abriss Schwechten-Halle, vor der letzten Modernisierung 1988
Moabit I, Schwechten hall demolition, before final modernisation in 1988

Zwei Modellfotos, Zustand vor und nach der letzten Modernisierung 1988
Two model photos, condition before and after final modernisation in 1988

Die letzte Modernisierungsphase

Die jüngste Entwicklungsphase auf dem Kraftwerksstandort setzte infolge der verschärften Umweltschutzbestimmungen in den 1980er Jahren ein.

Mit der Einführung der Wirbelschichtfeuerung wurde die gesamte Kraftwerksanlage grundlegend erneuert. Die von Walter Henn entworfenen voluminösen Neuanlagen dominieren seit 1988 weithin die Stadtlandschaft am Westhafen und haben das Kraftwerksensemble, insbesondere mit dem Maschinen- und Kesselhaus, in Gestalt und Dimension in einen neuen Maßstab überführt. Moabit I musste dieser Neukonzeption fast komplett weichen, lediglich der Kopfbau der alten Maschinenhalle blieb erhalten. Die Nordfassade wurde originalgetreu rekonstruiert. Die Gasturbinen wurden schrittweise an den Ostrand des Grundstückes verlegt.

Die drei Heizblöcke des alten Werkes II sind seit 1992 komplett außer Betrieb. Die neue Anlage erreichte 1989 mit 100 MW ihre volle Leistung und wird damit der Stadt Berlin langfristig für die Strom- und Wärmeversorgung zur Verfügung stehen.

The Last Modernisation Phase

The most recent development phase on the power station's site began as a result of stricter regulations for environmental protection in the 1980s.

The introduction of fan-assisted firing led to the renewal of the entire power station installation. Since 1988, the new, high-volume works, designed by Walter Henn, largely dominates the city landscape at Westhafen. Furthermore, the new works, especially the machine and boiler-house, have projected the power station ensemble into a new dimension in terms of structure and scale. Moabit I was almost completely transformed by this new design, with only the head structure of the old machine hall remaining intact. The northern façade was reconstructed according to original designs. The gas turbines were progressively moved to the eastern edge of the site.

The three heating blocks of Installation II were completely taken out of service in 1992. The new installation reached its full capacity of 100 MW in 1989 and this means that it is available on a long-term basis to the city of Berlin for electricity and thermal energy supply.

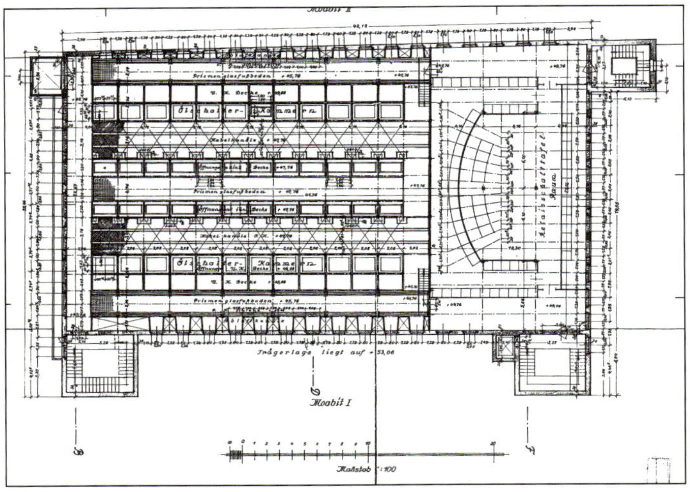

Riegelbau mit 30-kV-Schaltanlage, Grundriss, Zustand um 1957
Half-timber construction with 30 kV switching plant, ground plan, condition around 1957

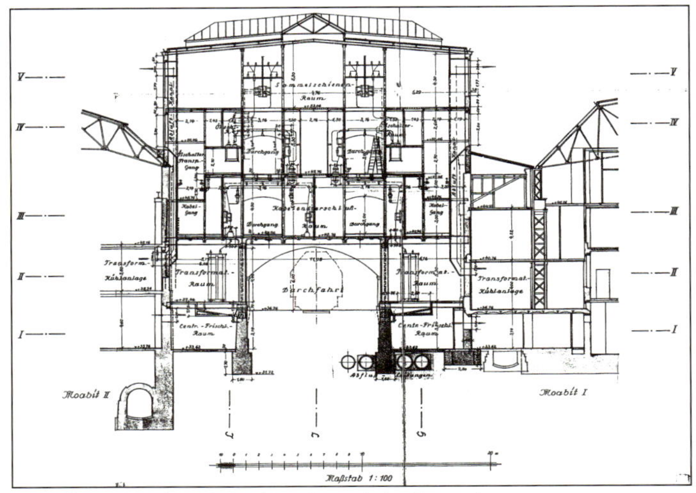

Riegelbau mit 30-kV-Schaltanlage, Querschnitt, Zustand um 1957
Half-timber construction with 30 kV switching plant, cross-section, condition around 1957

Das Denkmalensemble

Das Kraftwerk Moabit liegt als kulturhistorisches Zeugnis an der Schnittstelle des ausgehenden 19. Jahrhunderts und des beginnenden modernen, durch die Industrie geprägten Zeitalters. Für Franz Schwechten war Moabit einer seiner letzten Industriebauten. Bei der äußeren Gestaltung musste der Architekt eindrucksvoller historischer Prachtentfaltung von seinen traditionellen, akademischen Grundsätzen abrücken. Das Werk Moabit markierte den letzten nur noch leicht historisierenden, nicht mehr „historistischen" Industriebau Berlins.

Die nördliche, zum Kanalufer gerichtete Hauptfassade wurde beherrscht von dem leicht barockisierenden Giebel der ehemaligen Maschinenhalle mit einem mächtigen, durch zarte Stahlsprossen gegliederten Rundbogenfenster und sparsam angewandter Baukeramik. Der Giebel wurde auf der Ostseite flankiert von dem bescheideneren Anbau des Kesselhauses, auf der Westseite ragte ein gewaltiger quadratischer Turm mit achteckigem Obergeschoss und spitzer Haube auf, dem sich ein überraschend kleines Portal anschloss. Moabit I muss als der erste Teil einer Anlage angesehen werden, deren Erweiterung bereits vor Baubeginn so gut wie sicher war und die mit einer spiegelbildlichen Wiederholung eine große Fassadenkomposition hätte entstehen lassen. Die Erweiterung von Moabit II war dann tatsächlich eine spiegelbildliche Verdoppelung des Vorhandenen, allerdings nicht detailgetreu und in den Abmessungen durch den Einsatz von Turbinen

The Listed Building Ensemble

Moabit Power Station is a cultural and historical record of the transition from the end of the 19th century to the beginning of the industrialised, modern era. For Franz Schwechten, Moabit was one of his last industrial buildings. The architect had to distance himself from traditional, academic principles in favour of impressive and historical grandeur for the exterior design. The Moabit plant marked the last of Berlin's industrial constructions that was only slightly historicising but no longer "historical".

The northern main façade, built towards the canal-side, was dominated by the slight baroque element of the gable on the former machine hall, with an imposing round-arch window, formed by light, steel struts and carefully applied building ceramics. The gable was flanked on the eastern side with a modest extension to the boiler-house. On the western side, a massive square tower was errected, with an octagonal upper-storey and pointed roof, joined by a surprisingly small portal. Moabit I must be seen as the first part of an installation that was almost certainly to be extended before it was built and to have created a façade composition with mirror-imaging repetition. The extension of Moabit II was then a real mirror image, which doubled what already existed, although it was not an exact replica of detail and was on a scale that was for more modest due to the operation of turbines. The design of halls built parallel to each other was adopted. Schwechten arranged the series of wide

Moabit I, Schaufront zum Kanal, historische Zeichnung
Moabit I, Show-front towards the canal, historical drawing

Anlage 1989, mit Schaufront von Moabit I,
im Hintergrund Neubauten von W. Henn
Site in 1989, with show-front of Moabit I,
in the background, new buildings by W. Henn

weitaus bescheidener. Übernommen wurde das Konzept parallel angeordneter Hallen. Schwechten arrangierte die Folge von breitem Kesselhaus und schmaler Maschinenhalle zu einem symmetrischen Bild, das, durch die Erweiterungen der Folgezeit überformt, heute nicht mehr wahrnehmbar ist.
Im Gegensatz zu der architektonisch ausformulierten Fassade zeigte sich das Halleninnere zurückhaltend, dem Zweck Rechnung tragend. 117 Meter lang und 27 Meter breit beeindruckte die Halle allein schon durch ihre Größe und technische Ausstattung. Vorbild war das 1897 von der AEG errichtete Drehstromkraftwerk, die Zentrale Oberspree. In der Maschinenhalle fand sich die gleiche konstruktive Lösung für die Dachbinder und die Lichtführung, das gleiche Prinzip von massiver Mauerwerkswand auf der einen und einer Eisenfachwerkwand auf der anderen Längsseite mit einer auf halber Höhe angeordneten Galerie.
Im Gegensatz zu dem kurz zuvor fertig gestellten Kraftwerk Charlottenburg war die Formensprache der Maschinenhalle Moabit I von nüchterner Sachlichkeit geprägt. Die weißen Glasursteine der Wände mit ihren sparsamen blauen Akzenten verstärkten den Eindruck profaner Kühle. Die Halle mutete in ihrer Klarheit wie ein Vorgriff auf den nach 1900 einsetzenden modernen Industriebau an.

boiler-house and narrow machine hall to a symmetrical image that can no longer be detected today, due to the re-modeling of extensions over successive years.
By contrast to the architecturally designed façade, the interior of the hall-structure appeared rather restrained, taking its function into account. The hall's length of 117 metres and its width of 27 metres was impressive, purely in terms of size and technical equipment. The model was the alternating current power station, Central Oberspree, built in 1897 by AEG. In the machine hall, the same constructive solution was adopted for the roof beams and lighting and the same principle of massive stonework on one of the long sides and on the other an iron latticework wall with a gallery situated halfway up the wall.
By comparison to the Charlottenburg Power Station that was finished shortly beforehand, the formal style of Moabit I's machine hall was defined by sober objectivity. The white glass stones of the walls with their sparse blue tones heightened the impression of secular restraint. The hall's clarity suggested an early example of modern industrial construction, beginning after 1900.

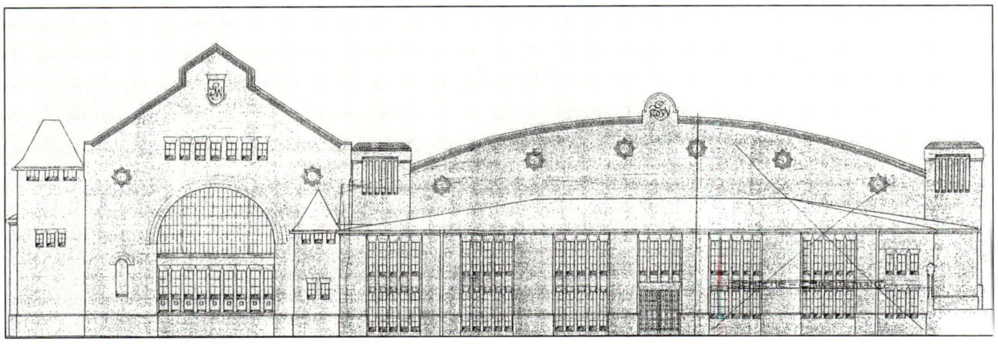

Moabit II, Nordansicht von Schwechten,
historische Zeichnung, 1908
Moabit II, northern view by Schwechten,
historical drawing 1908

Moabit I, Maschinenhalle, hist. Aufnahme 1907
Moabit I, machine hall, historical photograph 1907

Moabit I, Fragment der alten Maschinenhalle, Aufnahme 1986
Moabit I, fragment of old machine hall, photograph 1986

Die Erweiterungsbauten der 1920er Jahre

Neben dem historischen Relikt Moabit I haben die Erweiterungsbauten aus den 1920er Jahren das Denkmalensemble des Kraftwerks architektonisch geprägt. Das Schalthaus, als Brückenbau über dem ehemaligen Werkshof zwischen Moabit I und II errichtet, zeigt eine Ziegelfassade mit expressionistischen Details. Die regelmäßige kleinteilige Fensterordnung teilt wenig vom eigentlichen Zweck des Gebäudes mit. Hinter dem vorgeblendeten Bogenmotiv des Durchgangs mit sorgfältig detaillierter Ziegelverarbeitung wird die Durchfahrt durch eine rechtwinklige, große Stahlträgerordnung überfangen.
Eine sachlichere, dennoch expressive Sprache kennzeichnete die Kohlenstaubmahlanlage. Dieser 1925 ausgeführte, wohl proportionierte kubische Baukörper trug mit seiner betont vertikalen Gliederung der Schaufassaden zur Putlitzbrücke und zum Bahngelände, dem Kranzgesims und darüber ausladenden Walmdach deutlich die Handschrift von Hans Müller, dem damaligen Chefarchitekten der Bewag-Bauabteilung. Das historische Erscheinungsbild ging durch den Umbau in den 1950er Jahren, besonders aber durch die dreigeschossige Aufstockung in den 1980er Jahren, der letzten Ausbauphase des Kraftwerks, verloren.

The 1920s Extension Buildings

Alongside the historical relic of Moabit I, the extended construction work during the 1920s architecturally defined the ensemble of listed power station buildings. The switching house, erected as a bridge construction, between Moabit I and II, over the work's courtyard, displays a brick façade with expressionist details. The regular, small-scale window design reveals little about the building's actual function. Behind the forward placing of the arch-motif on the passageway, with careful detailing on the brick-work, the passage is enclosed by a right-angled, large, steel-girder construction.
The coal-dust grinding plant is characterised by more objective, yet still quite expressive formal style. This well-proportioned, cubic construction, built in 1925, as well as emphasis on vertical structuring of the main façades towards Putliz bridge and the railway site, the roof-ledge and the wide hipped roof above, clearly bore the signature of Hans Müller, then the chief architect of the Bewag's building department. The historical appearance was lost by reconstruction in the 1950s, especial-

30-kV-Schalthaus, Südseite, Aufnahme 1929
30 kV Switching house, south side, photograph 1929

Umformwerk von Klingenberg/Issel, Nordfront zum Kanal,
Aufnahme aus den 1950er Jahren
Transformer work by Klingenberg/Issel, north front towards the canal,
photograph from the 1950s

Den Abschluss der baulichen Maßnahmen stellte der Umbau von Moabit II mit dem Umformerwerk dar, der in den Jahren 1929/30 von den Architekten Waltar Klingenberg und Werner Issel durchgeführt wurde. Das vorhandene Werk wurde durch eine neue Fassadengestaltung überformt, wodurch auch die ursprüngliche Giebelfassade von Franz Schwechten verloren ging. Die gewählten formalen Mittel zur Gliederung der langen Fassaden bestanden in der rhythmisierenden Wiederholung eines abstrahierten klassizistischen Portikusmotivs. Als einfache gliedernde Ordnung verband es die verschiedenen Baukörper zu einer formalen Einheit.

ly by the addition of three additional storeys in the 1990s, the power station's last development phase.

The rebuilding of Moabit II with the converter plant, built in 1929/30 by architects Waltar Klingenberg and Werner Issel, marked the completion of building measures. The existing installation was re-modeled with a new façade structure. In the process, the original gable façade by Franz Schwechten was lost. The chosen formal means to structure the long façade was the rhythmic representation of an abstract, classical-style portico motif. This simple, structured pattern joined the various building objects into a formal unity.

Kohlenstaubaufbereitung von Hans Müller, Aufnahme
von der Putlitzbrücke nach der Errichtung 1925
Coal-dust processing by Hans Müller, photograph after
construction in 1925, view from Putlitz bridge

Kraftwerk Moabit • Moabit Power Station Friedrich-Krause-Ufer 10-15, 13353 Berlin	
Grundstücksfläche gesamt Land Area, in total	52.714 m²
Vermarktbare Flächen **Marketable Sites**	
Grundstücksfläche Land Area	4.500 m²
Bruttogeschossfläche (BGF) gesamt Gross Floor Area, in total (BGF)	9.400 m²
mit: / with: Maschinenhalle Machine Hall	3.300 m²
Kesselhaus Boiler-house	4.900 m²

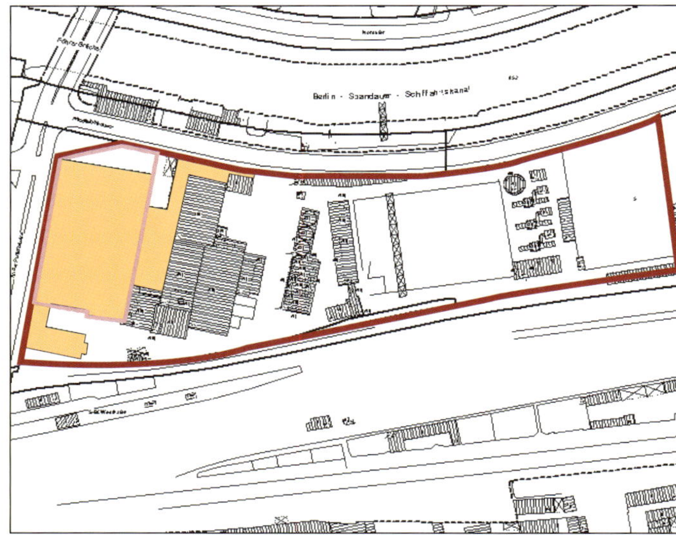

vermarktbare Grundstücksfläche — Marketable Land Area
Denkmalschutz — Listed Building Preservation

Zustand und Nutzung im Jahr 2003

Die erhaltenen und rekonstruierten historischen Bauteile von Moabit I und das Werk II mit dem Schalthaus und der Überformung der 20er Jahre stehen heute unter Denkmalschutz. Im Zusammenspiel mit den voluminösen Neubauten der 1980er Jahre dokumentiert das Ensemble in herausragender Weise die Geschichte des einst größten Berliner Kraftwerks.
Mit dem von Walter Henn konzipierten Neubau der gesamten Kraftwerksanlage erfolgte 1987 der Teilabriss des alten Kessel- und Maschinenhauses von Moabit I. Erhalten blieb die Maschinenhalle hinter der Fassade in einer Tiefe von zehn Metern sowie der dreigeschossige westliche Bürotrakt. Die historische Hauptfront, die durch Umbauten der Nazi-Zeit und Kriegszerstörungen stark beeinträchtigt war, wurde dabei originalgetreu rekonstruiert. Das große Rundbogenfenster im Hallengiebel, das vermauert war, wurde wieder hergestellt, der Helm am Eckturm des Eingangsbereiches wieder aufgebaut, die Tordurchfahrt und ein Teil der alten Umfassungsmauer nach alten Plänen wieder in ihren ursprünglichen Zustand versetzt. Dieser rekonstruierte historische Teil stellt heute das einzige bauliche Vermächtnis der ursprünglichen Kraftwerksanlage von Franz Schwechten dar. Das Fragment der alten Maschinenhalle wird durch die Bewag als repräsentativer Ausstellungsraum genutzt. Das Werk Moabit II ist in seinem äußeren Erscheinungsbild mit den Erweiterungen und Überformungen der 1920er Jahre weitgehend erhalten. Nach der letzten Moderni-

Condition and Usage in 2003

The preserved and reconstructed historical building sections of Moabit I and Plant II, with the switching house and the re-modeling of the 1920s, are today classified as listed buildings. The ensemble is an outstanding document, – in combination with the new high-volume 1980s buildings – of the history of what used to be one of Berlin's major power stations. In 1987, Walter Henn's newly designed construction of the entire power station site led to the partial demolition of the old boiler-house and machine shop of Moabit I. The machine hall, located behind the façade to a depth of ten metres, and the three-storey western office-wing, were preserved. The process included reconstruction work, according to the original plans, of the historical main frontage that was badly affected by rebuilding in the Nazi era and destruction during the war. The great round-arch window in the brickwork gable of the hall was returned to its original state, the spire on the corner tower of the entrance-area was rebuilt, the gateway and passage and part of the old perimeter wall were put into their original state, according to old plans. Today, this reconstructed, historical part represents the only building legacy of the original power station plant by Franz Schwechten. The fragment of the old machine hall is used by the Bewag as a show-case exhibition hall. The Moabit II plant is largely preserved in its exterior appearance with the extensions and re-modeling work of the 1920s. After the last wave of modernisation, the buildings have been reprieved from their original

Moabit I, nördliche Schaufassade,
Rundbogenfenster nach Wiederherstellung 1989
Moabit I, northern show façade,
round-arch window after renovation in 1989

Moabit I, Ausstellungsraum im Fragment der alten Maschinenhalle, 1989
Moabit I, exhibition room in fragment of old machine hall, 1989

Kraftwerksensemble heute, von Putlitzbrücke und Kanal
Power station ensemble today, view from Putlitz bridge and canal

sierung sind die Gebäude ihrer ursprünglichen Funktion enthoben und stehen dem Kraftwerksbetrieb nur noch für untergeordnete Nutzungen zur Verfügung. Lediglich der nördliche Kopfbau des Brückengebäudes ist ausgebaut und wird als Geschossbau von der Kraftwerksverwaltung genutzt. Im alten Kesselhaus II ist noch die technische Ausrüstung der Nachkriegszeit vorhanden. Das Denkmalensemble von Moabit II wird mittelfristig für neue Nutzungen zur Verfügung stehen.

Die planungsrechtliche Situation

Planungsrechtlich ist der Baunutzungsplan von 1960 maßgeblich, der den Kraftwerksstandort als reines Arbeitsgebiet mit Baustufe 6 kennzeichnet. Der Flächennutzungsplan von 2000 weist den Standort als Fläche mit gewerblichem Charakter, Zweckbestimmung Energie aus. Bei einer Umnutzung des Denkmalensembles ist von einer Modifizierung der verbindlichen Bauleitplanung auszugehen.
Entlang des Kanals ist ein durchgängiger öffentlicher Uferwanderweg vorgesehen, der vom Humboldt-Hafen über Nord- und Westhafen eine Verbindung vom nördlichen Tiergarten mit Regierungsviertel bis zum Volkspark Rehberge schafft.

Nutzungs- und Entwicklungspotenziale

Mit der großen Modernisierung und Neukonzeption des Kraftwerks in den 1980er Jahren steht das alte Werk Moabit II mit den Erweiterungsbauten der 20er Jahre vor einer grundlegenden Umorientierung. Für das denkmalgeschützte Ensemble werden mittelfristig neue Perspektiven als Überlebensstrategie zu entwickeln sein. Unterstützend wird die überarbeitete Bereichsentwicklungsplanung wirken, die mit einem ausgedehnten Ufergrünzug die Lage am Berlin-Spandauer-Schifffahrtskanal der Öffentlichkeit erschließen und aufwerten wird. Die städtebaulich markante Situation des Ensembles an der Putlitzbrücke und zum Kanal orientiert eröffnet ein breites Spektrum von Nachnutzungsoptionen in den Bereichen Sport, Freizeit sowie Gastronomie, Kunst, Kultur und Bildung.
Die räumliche und architektonische Qualität der alten Hallen und ihr im Stadtbild Identität stiftender Charakter werden wesentlich dazu beitragen, den entstandenen Freiraum mit neuen Funktionen wiederzubeleben.

function and are only available for minor use for the power station's operations. Only the northern head structure of the bridge complex is developed and used as the ground-floor building of the power station's administrative office. In the old boiler-house II, the technical equipment, dating from the post-war period, is still available. In the medium-term, the listed building ensemble of Moabit II will be available for new uses.

The Legal Planning Situation

The 1960s building utilisation plan sets the standard in legal planning terms, defining the power station site as a purely commercial area with building level six. The land utilisation plan, dating from 2000, defines the location as a plot with commercial character for energy purposes. Re-use of the listed building ensemble is likely to modify the legal planning of the site management.
Along the canal, a public canal-path is planned as a hiking route that connects from the Humboldt Harbour, via North and Westhafen and from the northern Tiergarten with government quarters to the Rehberge Volkspark.

Utilisation and Development Potential

The extensive modernisation and new power station design in the 1980s means that the old plant, Moabit II, with the 1920s building extensions, is about to undergo a basic overhaul. In the medium term, as a survival strategy, new perspectives will need to be developed for the listed building ensemble. The revised area development planning will play a supporting role, with an extended green belt that is to develop and improve the area for the public on the Berlin-Spandau shipping canal. The prominent, urban situation of the ensemble on the Putlitz bridge and towards the canal area opens a broad spectrum of re-utilisation options in areas of sport, leisure, catering, art, culture and education.
The spatial and architectural quality of the old halls and their distinctive character that gives the city landscape a unique identity will make an essential contribution to re-vitalising the developed, open space with new functions.

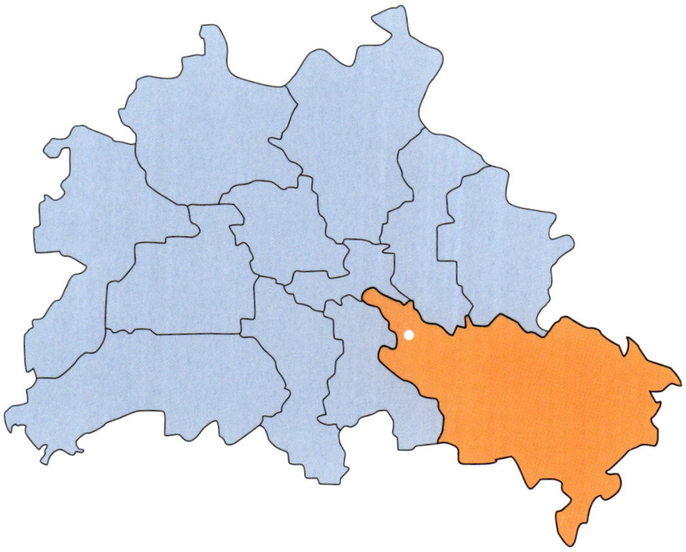

Kraftwerk Rummelsburg
Rummelsburg Power Station

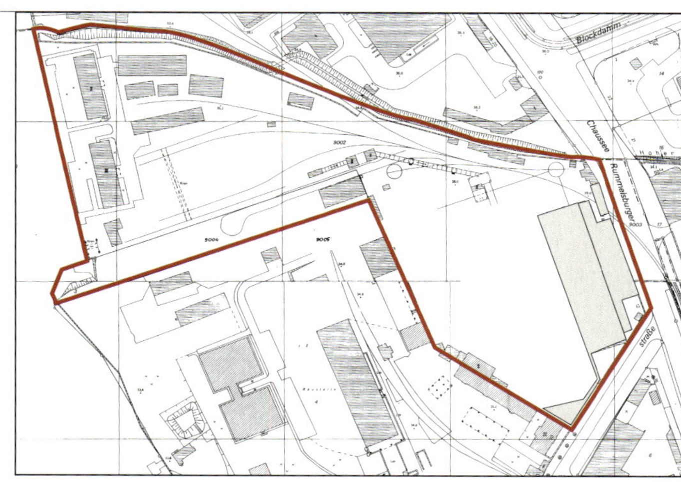

Lageplan
Site plan

DAS KRAFTWERK RUMMELSBURG
RUMMELSBURG POWER STATION
Rummelsburger Landstrasse 2/12, Nalepastrasse 2/8

Die stadträumliche Einbindung

Das Gelände des alten Kraftwerks Rummelsburg liegt in einem traditionsreichen Gewerbegebiet, das mit zu den größten in der Vergangenheit industriell genutzten Flächen in Rummelsburg gehört. Die Entwicklung des Industriegebiets Rummelsburg setzte um die Jahrhundertwende mit der Eröffnung eines Teils der Ringbahn und der Station Stralau-Rummelsburg, dem heutigen Ostkreuz, ein. Mit der direkten Anbindung an die Bahntrasse und begünstigt durch die unmittelbare Lage am Wasser war die Industrialisierung dieses Gebiets eng verbunden mit dem Ausbau der Energie- und Wärmeerzeugung.
Das alte Kraftwerk liegt auf einem unregelmäßig geschnittenen Grundstück zwischen der Rummelsburger Landstraße und der Spree in einem überaus heterogenen Umfeld. Nach Norden erstreckt sich der ausgedehnte Gewerbegürtel, der vom Kraftwerk Klingenberg weithin dominiert wird und seinen Abschluss in dem Entwicklungsgebiet „Rummelsburger Bucht" findet. Östlich schließt sich der Ortsteil Karlshorst mit teilweise villenartiger Bebauung aus dem beginnenden 20. Jahrhundert an, welche sich nach Süden in zusammenhängende Kleingartenanlagen auflöst. Im Westen bestimmen die Spree und der gegenüberliegende Plänterwald das Bild.
Das Kraftwerksensemble steht direkt an der Rummelsburger Landstraße, einer ehemaligen Chausseestraße, die seit Beginn des 20. Jahrhunderts durch anspruchsvolle Industriebauten geprägt wird und bis heute eine der großen Ausfallstraßen Berlins darstellt. Mit seiner Lage und monumentalen Gestaltung bildet das Kraftwerk Auftakt und Abschlusspunkt des traditionellen Industriebandes im Übergang zu der weitgehend offenen, von Kleingärten und Wald bestimmten Landschaft der Wuhlheide.

Urban Integration

The site of the former Rummelsburg Power Station is located on a traditionally commercial area that is amongst some of the largest, former sites for industrial usage in Rummelsburg. The development of the Rummelsburg industrial land began around the turn of the century with the opening of part of the ring railway network and the station Stralau-Rummelsburg, today known as Ostkreuz. With direct a connection to Bahnstraße and benefiting from the direct location on the water, the industrialisation of this area was closely connected to the development of energy and thermal energy production.
The old power station is located on irregularly defined land, between Rummelsburger Landstraße and the river Spree, in a highly heterogeneous surrounding area. Towards the north lies the extended commercial belt that is largely dominated by Klingenberg Power Station and ends with the development area of "Rummelsburg Bay". Towards the east, lies the district of Karlshorst, partly with its villa-style buildings from the early 20th century, ending in a series of allotments towards the south. In the west, the river Spree and the Plänter woods on the opposite side dominate the site.

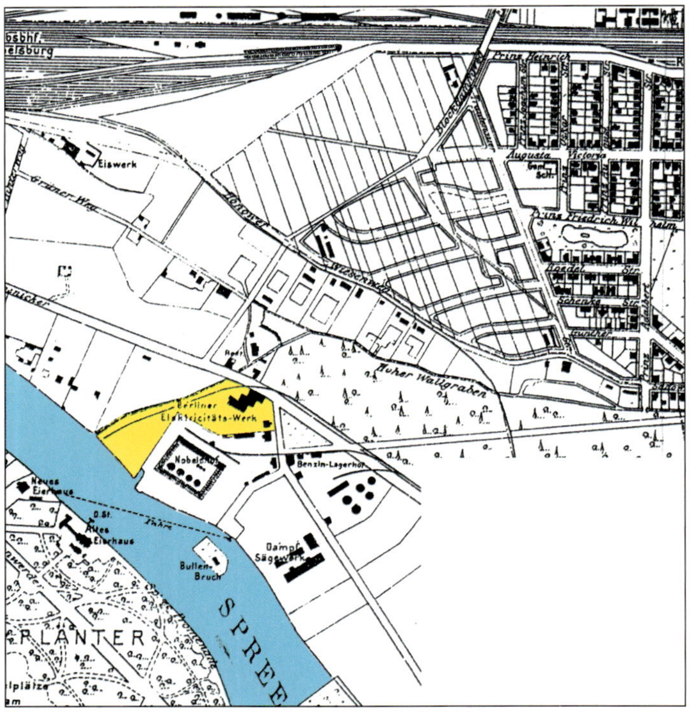

Übersichtsplan 1911
Overview plan 1911
1:15000

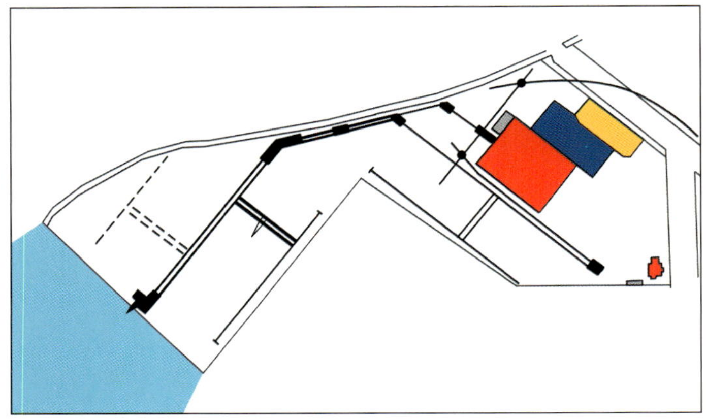

1907

Entwicklungsstufen des Kraftwerks
Power station's development in phases

Die Entwicklung des Kraftwerks

Vorgeschichte

Dem rasant fortschreitenden Bedarf an der neuen Energie, der Anfang des 20. Jahrhunderts mit der Industrialisierung und aufstrebenden Entwicklung Berlins zur Metropole einherging, konnten die Berliner Elektrizitätswerke (BEW) wegen der vertraglich festgelegten Ausbaubegrenzungen für die Kraftwerke bald nicht mehr begegnen. Im Jahr 1906 mussten Neuanträge auf Stromanschluss abgelehnt werden. Nach langwierigen Verhandlungen mit dem Magistrat wurden 1907 die Kapazitätsbegrenzungen aufgehoben und die Gesellschaft verpflichtet, „Elektrizität für Licht und sonstige, insbesondere auch Kraftzwecke zu liefern und die Werke dementsprechend zu erweitern und zwar in dem Umfange, dass sie jeglichen im Weichbilde von Berlin hervortretenden Bedürfnissen genügen".

Noch im gleichen Jahr wurde mit dem Bau des Kraftwerks Rummelsburg begonnen, nachdem das erste Drehstromkraftwerk Oberspree vollständig ausgebaut war und keine baulichen Erweiterungen mehr zuließ. Die geplante Erweiterung der Zentrale Moabit, Moabit II, musste für die Versorgung der westlichen Stadtteile vorbehalten bleiben. Das Kraftwerk Rummelsburg versorgte ab 1907 zusammen mit der Zentrale Oberspree die östliche Hälfte Berlins und die umliegenden Orte. Mit seiner Erstausstattung von drei Turbodynamos überstieg es mit einer Anfangsleistung von 13.500 kW bei weitem die Leistungskraft bisheriger Kraftwerke. Es war das erste Kraftwerk der BEW, in dem ausschließlich Turbinen verwendet wurden.

Gleichzeitig mit dem Bau des Kraftwerks Rummelburg wurde das älteste Berliner Kraftwerk in der Markgrafenstraße stillgelegt.

Das Grundstück

Das von der BEW gewählte und 1906 erworbene Grundstück wies durch den Eisenbahnanschluss und die direkte Wasserlage günstige Voraussetzungen für die Kühlwasserbeschaffung sowie den Kohlen- und Materialtransport auf. Mit einer kurzen Anbindung an die bereits ausgebaute Oberschöneweider Industriebahn in der Rummelsburger Landstraße war die Anbindung des Grundstücks perfekt.

Eine professionelle Planung sicherte die optimale Nutzung des unregelmäßigen Grundstücks: An der Spree wurden die Kohlen-

The power station ensemble stands directly on Rummelsburger Landstraße, a former thoroughfare, characterised by impressive industrial buildings from the beginning of the 20th century. To this day, this street is one of Berlin's major city routes. Due to its location and monumental structure, the power station is both the start and finish of the traditional industrial belt leading into the largely open landscape of the Wuhlheide, a heath area, dominated by allotments and woods.

The Development of the Power Station

Pre-History

The steep rise in the need for new energy that coincided, at the beginning of the 20th century, both with the industrialisation and ambitious development of Berlin into a metropolis, could soon no longer be met by the Berlin Electricity Works (BEW), due to contractual restrictions on power station capacity. In 1906, new applications for electricity connection had to be turned down. After tedious negotiations with the magistrate's office, capacity restrictions were suspended in 1907 and the society was under an obligation "to supply electricity for light and other, especially power-related purposes and to expand the plants accordingly and, to be precise, on a scale such as any needs arising in any sector of the urban area could be satisified."

In the very same year, the building of Rummelsburg Power Station was begun, after the first alternating current power station, Oberspree, was fully developed, thus permitting no further building extensions. The planned extension of Central Moabit, Moabit II, had to be reserved for the supply of the western sectors of the city. From 1907, Rummelsburg Power Station and Central Oberspree supplied the eastern half of Berlin and the surrounding areas. With its original equipment of three turbodynamos, it far exceeded the capacity of any previous power stations with a initial capacity of 13,500 kW. It was the BEW's first power station to run exclusively on turbines. At the same time as building work began on Rummelsburg Power Station, Berlin's oldest power station at Markgrafenstraße was closed down.

The Land

The land that was chosen by the BEW and acquired in 1906 indicated beneficial conditions for the supply of cooling water and for coal and material transportation by connection to the railway network and the location directly on the water. A short connection to the already developed Oberschönweider-industrial railway, at Rummelsburger Landstraße, meant that this side's transport connections were perfect. Professional planning secured the optimal usage of this irregular site:

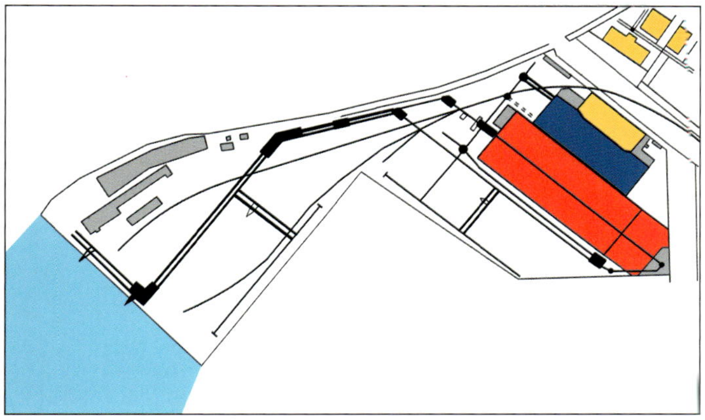

1926

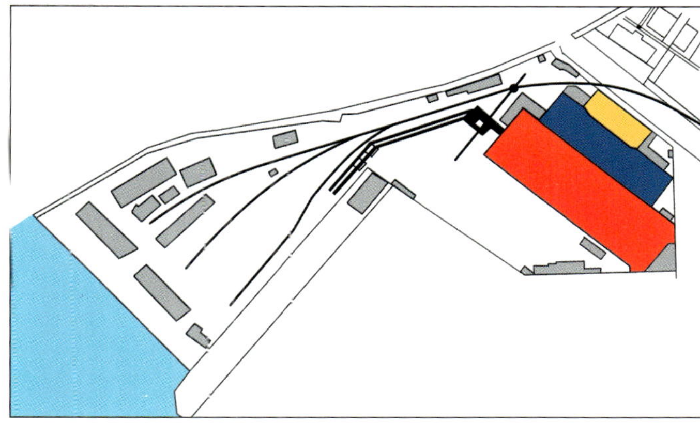

1975

Wasserstraßen	Waterways
Kesselhaus	Boiler-house
Maschinenhaus	Machine Hall
Stromverteilung	Electricity Distribution
Nebengebäude aller Art	Various Outbuildings
Kohlentransport	Coal Transport

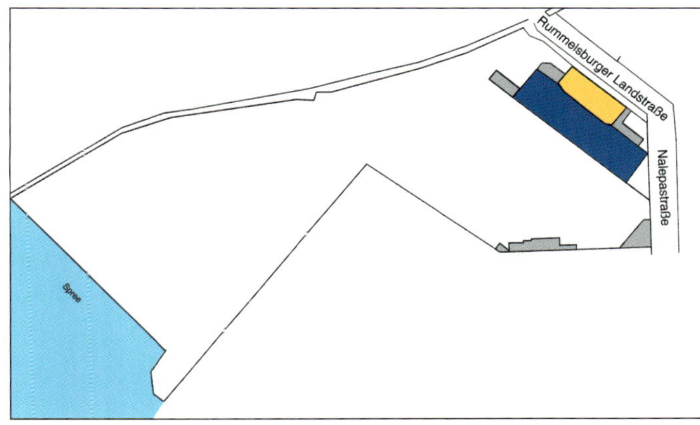

2001

lagerplätze angeordnet, die durch eine nach modernsten Gesichtspunkten angelegte Kohlenförderungsanlage mit den auf der Westseite des Grundstücks, unmittelbar an der Rummelsburger Landstraße liegenden Kraftwerksgebäuden verbunden waren.

Die Anlage von 1907

Die Baugeschichte des Kraftwerks begann 1907 mit dem Abriss eines Chausseehauses. Als Ersatz sollte auf der Südseite des Grundstücks ein zweigeschossiges Beamtenwohnhaus entstehen. Der Entwurf der Kraftwerksneubauten stammte aus dem Baubüro der BEW, ein verantwortlicher Architekt ist nicht benannt. Sowohl eine nicht realisierte Entwurfsfassung als auch die ausgeführte Anlage zeigen deutlich, dass es sich um eine vereinfachte, moderne Version des Kraftwerks Oberspree handelt. Das Werk Rummelsburg wurde in Parallelanordnung von Kesselhaus und Maschinenhalle mit dem straßenseitigen Anbau eines Schalt- und Betriebsgebäudes konzipiert.

Der steigende Strombedarf bestimmte den schrittweisen Ausbau der Anlage. Der südliche Teil der beiden Hallen blieb zunächst unausgeführt, erst 1915 war das Kraftwerk seiner ursprünglichen Planung entsprechend vollendet. Erweiterungen in den Kriegsjahren ließen die Kraftwerksleistung bis Anfang der 20er Jahre auf 52.000 kW ansteigen.

coal storage facilities were located on the Spree, connected to the power station buildings on the western side of the plot, at Rummelsburger Landstraße, by the most modern equipped coal transportation plant.

The Installation in 1907

The power station's building history began in 1907 with the demolition of one of the buildings on the main thoroughfare. On the southern side of the plot, a two-storey accommodation block for officials was to be built as a replacement. The design of the power station's new buildings emerged from the BEW's building office, no responsible architect being named. A draft version of the design and the finished installation clearly shows that the resulting building was a simplified, modern version of Oberspree Power Station. The Rummelsburg plant was designed with boiler-house and machine halls built parallel to each other with the building of a switching and operational unit towards the street.

The increasing demand for electricity determined the plant's gradual development. Initially, the southern part of both halls was not developed. The power station was only completed to its original plan in 1915. Extensions during the war years enabled the power station's capacity to climb to 52,000 kW in the 1920s.

Förderanlagen an der Spree, Aufnahme um 1911
Conveying units on the Spree, photograph around 1911

Maschinenhalle, Südliche Erweiterung von Hans Müller, im Bau
Machine hall, southern extension by Hans Müller, under construction

Maschinenhalle mit Erweiterung von Hans Müller, Innenaufnahme um 1930
Machine hall with extension by Hans Müller, interior view around 1930

Erweiterungen der 1920er Jahre

Das neue Bauprogramm, das die Bewag in den 1920er Jahren anlässlich der Bildung von Groß-Berlin auflegte, war der Auftakt für eine zweite große Ausbauphase im Kraftwerk. Unter der Leitung des Architekten Hans Müller wurden von 1925 bis 1929 Kesselhaus und Maschinenhalle bis an die südliche Grundstücksgrenze verlängert, so dass das kleine Beamtenwohnhaus verloren ging. Nach dieser Erweiterung war die geplante Kapazität mit acht Turbinensätzen und einer Leistung von 79.600 kW erreicht, die Baumaßnahmen für das Kraftwerk Rummelsburg im Wesentlichen abgeschlossen.

Kriegsjahre und Nachkriegszeit bis zur Stilllegung 1966

Um 1918 war mit der Einspeisung von Fernstrom aus dem Braunkohlengroßkraftwerk Golpa-Zschornewitz bei Bitterfeld über eine 132 Kilometer lange 100-kV-Freileitung ins Berliner Netz begonnen worden. Dies führte zusammen mit der Inbetriebnahme des Großkraftwerks Klingenberg 1926 zu einer Entlastung der Stromversorgung. Da der Stromverbrauch der Industrie durch die Weltwirtschaftskrise außerdem merklich zurückgegangen war, konnte das Kraftwerk Rummelsburg 1930 vorläufig außer Betrieb genommen werden. Mit Beginn der Aufrüstung und während des Zweiten Weltkriegs wurden die Anlagen wieder reaktiviert. Da die Gebäude und technischen Einrichtungen im Krieg nahezu unversehrt blieben, konnte Rummelsburg nach 1945 als erstes Kraftwerk der Stadt wieder Strom ins Berliner Netz liefern. 1966 hatte das Kraftwerk nach 60 Jahren Betriebszeit ausgedient, die Stromerzeugung wurde endgültig eingestellt.

Zwischennutzungen und der Teilabriss des Kesselhauses

Die Folgejahre brachten Abbrucharbeiten im großen Stil mit sich. Die Außenanlagen mit den Förderbändern wurden abgetragen, Maschinen- und Kesselhaus ausgeräumt, die Schornsteine umgelegt. Bis zur Wende standen Teilbereiche der alten Kraftwerksgebäude noch für untergeordnete Nutzungen zur Verfügung. Nachdem jedoch durch fortschreitenden Verfall, insbesondere des Kesselhauses, inzwischen Gefahrensituationen entstanden waren, konnte Ende der 1990er Jahre eine Teilabrissgenehmigung für das Kesselhaus erwirkt werden. Bis 2001 war der baufällige Teil hinter der Müller'schen Fassade abgerissen.

The 1920s Extensions

The new building programme that the Bewag produced in the 1920s, with the creation of Greater Berlin, was the signal for a second, major wave of building work in the power station. From 1925 to 1929, under the directorship of the architect Hans Müller, the boiler-house and machine hall were lengthened as far as the southern perimeter of the site, so that the small accommodation unit for officials was lost. After this extension, the planned capacity of eight turbine sets and an output of 79,000 kW was reached. The building measures for Rummelsburg Power Station were essentially finished.

The War Years and Post-War Era until the Closure in 1966

Around 1918, the change to long-distance energy from the high-capacity brown coal power station, Golpa-Zschornewitz, near Bitterfeld, was begun with a 132 kilometre, 100 kV overhead transmission line into Berlin's grid. As a result, along with the operation of the high capacity Klingenberg Power Station in 1926, pressure on Berlin's electricity production was reduced. Industrial energy consumption had markedly decreased, due to the world economic crisis and, in 1930, Rummelsburg Power Station could be completely taken out of service. The beginning of re-armament and World War Two meant that both plants were operational again. The building and technical installations were virtually untouched during the war so that, after 1945, Rummelsburg was the

Abriss Kesselhaus hinter der Müller´schen Fassade, Aufnahme 2001
Demolition of boiler-house behind Müller's façade, photograph 2001

Historische Anlage, Luftaufnahme von Norden, um 1910
Historical site, aerial photograph from the north, around 1910

Das Denkmalensemble

Das 1907 errichtete Kraftwerk Rummelsburg dokumentiert als erstes modernes Dampfturbinenwerk in Berlin den enormen technischen Fortschritt vor dem Ersten Weltkrieg im Wechselspiel mit dem Ausbau Berlins zur Metropole. Es verkörpert als ein frühes Beispiel den Prototyp des jungen modernen Hochspannungskraftwerks, der Ausgangspunkt für den Sieg der AEG-Kraftwerksbauten in der ganzen Welt war. Dem Selbstbewusstsein der aufstrebenden Elektroindustrie entsprach eine sachliche, dem Zweck Rechnung tragende, aber durchaus anspruchsvolle Formensprache. Das Ensemble verkörpert eine von historisierender Ornamentik gereinigte Architektur, die in vielen Details für die damalige Zeit moderne und überzeugende Antworten auf die Bauaufgabe fand.
Die historische Anlage zeigte in ihrer Vollständigkeit eine Hintereinanderstaffelung von einem flächenmäßig großen, zur Spree orientierten Kesselhaus, der hohen, im Zentrum stehenden Maschinenhalle und dem zur Straße hin repräsentativ ausgerichteten Anbau der Schaltanlage mit Verwaltungstrakt.
Das ehemalige Kesselhaus, eine breit gelagerte, zweischiffige Halle mit mittig angeordnetem Kohlenbunker, bildete den rückwärtigen Abschluss des Ensembles. Überspannt wurde die Halle durch eine fein-

Kesselhaus, historischer Schnitt
Boiler-house, historical section

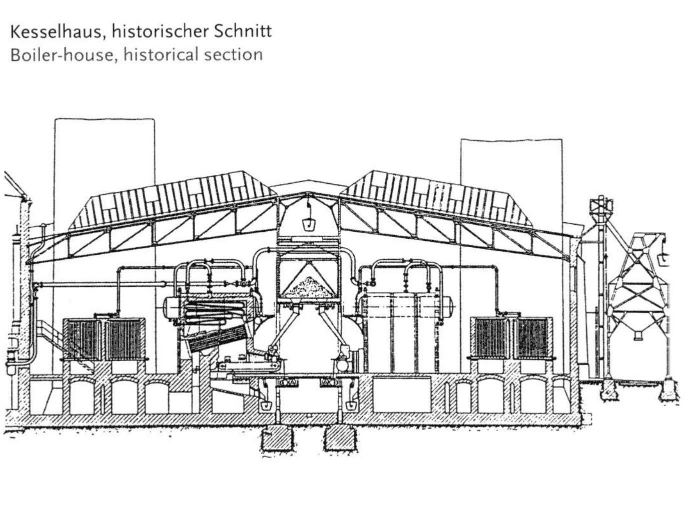

city's first power station to supply electricity to Berlin's grid. In 1966, after sixty years' service, the power station had served its time and electricity production was finally stopped.

Interim Usage and Partial Demolition of the Boiler-House
The following years led to large-scale demolitions. The external plants with the conveyor belts were dismantled, the machine and boiler-houses were cleared and the chimneys were demolished. Until German unification, partial sectors of the former power station buildings were available for minor usage. Meanwhile, however, after rapid deterioration, especially to the boiler-house, dangerous situations had emerged and at the end of the 1990s, permission was obtained for a partial demolition of the boiler-house. Until 2001, the unstable section of the building, behind Müller's original façade, was demolished.

The Listed Building Ensemble

Rummelsburg Power Station, built in 1907, documents the first, modern steam turbine plant in Berlin and the enormous technical progress before World War One, in combination with Berlin's transformation into a metropolis. As an early example of a prototype model of one of the first, modern high-voltage power stations, it embodies the beginning of the victory of AEG's power station constructions worldwide. The ambitious confidence in the electro-industry was matched by an objective formal style that took into account function, and yet was highly sophisticated. The ensemble stands for a style of architecture, free of all historicising ornamentation, providing modern and convincing answers to the construction task in many details that were typical for the time.
In its completed state, the historical plant was a successive, staggered arrangement of large volume boiler-house standing towards the Spree, the high machine hall standing in the centre and the extension of the switching house built towards the street with an imposingly sited extension and administrative wing.
The former boiler-house, a broad-based, double-transept hall with a coal-bunker set in the middle, completed the ensemble towards the rear. The hall was covered by a finely structured iron-latticework con-

Kesselhaus Längsfassade, Aufnahme 1998
Boiler-house, lengthways façade, photograph 1998

gliedrige Eisenfachwerkkonstruktion mit aufgesetzten Oberlichtern. Die Kesselhausfassade aus rotem Ziegel war auf der Längsseite durch Strebepfeiler und Lisenen in überschaubare Flächen mit dreiteiligen Fensterelementen gegliedert, wodurch der stark horizontal ausgerichtete Bau eine ausgewogene Proportion erhielt. Im Nordgiebel der Halle wurden die einfachen Gliederungselemente der daneben liegenden Maschinenhalle aufgegriffen, in seiner dekorativen Ausgestaltung ordnete er sich dieser jedoch deutlich unter.

Maschinenhaus und Anbau in der Anlage von 1907

Als Kern und Herzstück der Anlage überragte die zweigeschossige Halle des Maschinenhauses mit einer Höhe von 25 Metern das gesamte Ensemble. Die Nordfront wird durch symmetrisch angeordnete große Segmentbogenfenster über einem leicht vortretenden Sockelgeschoss bestimmt. Lisenen und Eckrisalite, die bis zur schrägen Gesimsebene des Giebels aufragen, betonen die Dreiachsigkeit der Fassade. Auf den roten Ziegelflächen entfalten sparsam eingesetzte Ornamente eine fast grafische Wirkung.

Das Innere der Maschinenhalle lässt eine unübertreffliche Einfachheit und Übersichtlichkeit erkennen. Durch die hohe, filigrane Eisendachkonstruktion mit breiten Lichtbändern entsteht in der Halle ein lichtdurchfluteter Raumeindruck. Der äußersten Funktionalität steht eine ausgeprägte Ästhetik im Detail gegenüber, die sich in der Gestaltung von Treppen und Geländern, im Wechsel von roten und gelben Klinkerflächen und in medaillonartig eingelassenen Ziegelornamenten in den Wandflächen anspruchsvoll darstellt. Die rhythmische Gliederung der beiden Hallenlängswände durch gemauerte Pfeilervorlagen wird an der Ostseite in ihrer Abfolge durch die Schaltzentrale unterbrochen, die sich als gestalterischer Höhepunkt balkonartig in den Maschinensaal schiebt und gleichzeitig den Übergang zum Betriebsgebäude herstellt.

Der viergeschossige Bau greift die Fassadenarchitektur der Halle auf und überspielt so seine wahre Struktur und Zweckbestimmung. Weite Öffnungsachsen mit hoch aufragenden Rundbogenfenstern bestimmen die Schauseiten zur Straße. Die mit Pilastern sowie Vor- und Rücksprüngen stark plastisch gegliederte Fassade dokumentiert in ihrer Ausprägung mit Sockel-, Haupt- und Mezzaningeschoss den repräsentativen Anspruch und die Bedeutung der Anlage im Stadtbild.

struction with upper-lighting. The boiler-house façade, made of red brick, was structured on the long-side by buttresses and anchor blocks in manageable surfaces with a three-part window element, lending the construction's strong horizontal aspect a balanced proportion. In the hall's north gable, simple structuring elements were taken up from the adjacent machine hall, although the decorative effect was clearly below the standard of this structure.

Machine Hall and Extension in the 1907 Installation

The core and centrepiece of the installation was the two-storey hall of the machine shop that towered above the entire ensemble at a height of 25 metres. The northern frontage is characterised by great segment-arch windows positioned in symmetrical fashion above a slightly exposed basement-level. Anchor blocks and corner projections that stretch to the sloping ledge of the gable emphasise the three shafts of the façade. On the red brick areas, the sparse use of ornaments produce an almost graphic effect.

The interior of the machine hall reveals a unique simplicity and transparency. By means of filigree iron-roof construction with broad light tracts, the hall creates the impression of space that is flooded with light. In opposition to the extreme functionality, there is a marked and detailed aesthetic style that is expressed to a high standard in the design of stairways, landings, in the changing red and yellow clinker surfaces, and in the medal-shape of the brick ornaments let into the wall surfaces. On the eastern side, the central switching facility interrupts the development of the rhythmic structure of stonework pillars on each of the buildings' longer walls. The switching plant intervenes as the creative high point in a balcony-form in the machine hall and, at the same time, it forms a transition to the operations building.

The four-storey building adopts the façade-style architecture of the hall and so exaggerates its true structure and purpose. Wide opening shafts define the "best sides" to the street by high-reaching round-arch windows. The façade is made up of pilasters and projections and recesses in a strongly plastic effect and its definition with basement, main and mezzanine floor levels shows the imposing function and significance of the installation for the city image.

Betriebsgebäude, Ansicht Rummelsburger Landstraße, Aufnahme 2000
Operations building, view from Rummelsburger Landstraße, photograph 2000

Maschinenhalle mit Schaltzentrale, Aufnahme 1998
Machine hall with central switching plant, photograph 1998

Maschinenhalle, Treppe zur Schaltzentrale, 1998
Machine hall, stairway to central switching plant, 1998

Maschinenhalle mit Schaltzentrale, 1907
Machine hall with central switching plant, 1907

Die Erweiterungen durch Hans Müller in den Jahren von 1925 bis 1929

Von entscheidender architektonischer und städtebaulicher Qualität waren die zwischen 1925 und 1929 durch Hans Müller errichteten Erweiterungen.

Müller verlängerte die beiden Hallen soweit es das Grundstück zuließ und schloss die Maschinenhalle ihrer Bedeutung entsprechend mit einem Stufengiebel ab. Die durchgehende Halle zeigte sich am Außenbau durch schmale vertikale Fensterschlitze. Ein in die Südspitze des Kesselhauses eingepasster Stockwerksbau erhielt die dazugehörige Lochfassade. Das dazwischen liegende Treppenhaus teilte als vertikale Linie mit überhöhtem Rundbogenfenster die asymmetrisch gegliederten Kompartimente. Die glatten roten Ziegelwände und die durchgehende Silhouettenlinie des Stufengesimses binden die verschiedenen Rhythmen zusammen und schaffen eine harmonische Einheit. Mit ihrer monumentalen Expressivität dominiert die Südseite des Kraftwerk bis heute das Kraftwerksensemble und seine Umgebung.

The Extensions by Hans Müller from 1925 to 1929

The extensions built by Hans Müller between 1925 and 1929 were decisive in terms of quality both of architecture and urban construction.

Müller lengthened each of the halls, as long as the land permitted; and he completed the machine hall with a step-gable, in keeping with its significance. The extended hall showed itself on the outside by narrow, vertical window slits. The building of the levels, let into the south tip of the boiler-house, maintained the break in the adjoining façade. The stairwell lying in-between divided the asymmetrically structured compartments as a vertical line with raised round-arch windows. The smooth, red brick walls and the continual line of the silhouette of the stepped-ledge combine the various rhythms and create a harmonious unity. To this day, the southern side of the power station, with its monumental express veneness, dominates the ensemble and the surrounding area.

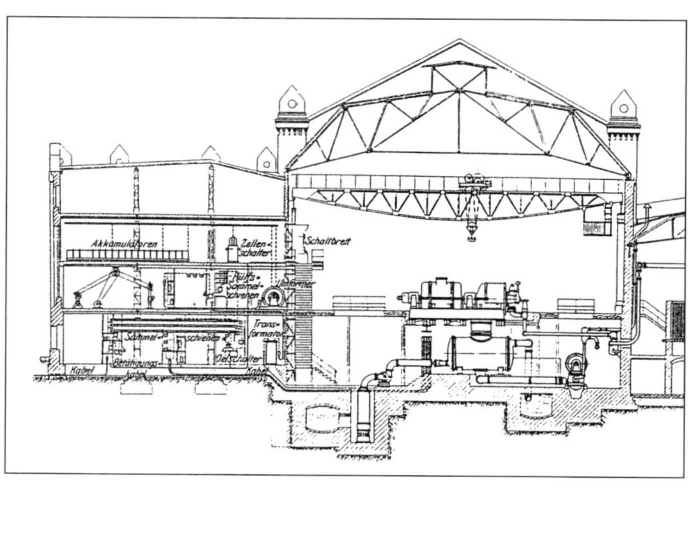

Maschinenhalle mit Betriebsgebäude, historischer Schnitt
Machine hall with operations buildings, historical section

Südseite mit der Erweiterung des Kessel- und Maschinenhauses von Hans Müller, Aufnahme 1992
South side with extension of boiler-house and machine hall by Hans Müller, photograph 1992

Zustand und Nutzung im Jahr 2003

Die Problematik des Umgangs mit dem alten Kraftwerk Rummelsburg wurde mit der endgültigen Stilllegung des Werks Ende der 1960er Jahre relevant. In den Folgejahren wurden die Hallen vom Ost-Berliner Energieversorger nur noch als Lagerstandort genutzt und die Bauunterhaltung aufgrund fehlender Nutzungsperspektiven für die Gebäude eingestellt. Fortschreitender Verfall führte 1986 zur Sperrung des Kesselhauses für den weiteren Betrieb.
1994 sah sich die Bewag veranlasst, der bestehenden Einsturzgefahr im Kesselhaus durch einen Antrag auf Abrissgenehmigung zu begegnen. Die Unterschutzstellung des Ensembles im gleichen Jahr führte zunächst zu kontroversen Standpunkten von Eigentümer und Behörde. Intensive Verhandlungen zwischen Bewag, Landesdenkmalamt und

Condition and Usage in 2003

The problem of using the old Rummelsburg Power Station was relevant at the end of the 1960s when the plant was finally closed down. In the following years, the halls were only used as a storage facility for energy suppliers to East Berlin and construction maintenance was stopped due to the lack of ideas for re-utilising the buildings. In 1986, progressive decline led to the closure of the boiler-house for further operation.
In 1994, the Bewag was obliged to respond to the imminent danger of collapse in the boiler-house by applying for a demolition permit. In the same year, the preservation status of the ensemble initially caused controversy between the owner and the authorities. Intensive negotiations between the Bewag, the federal office for listed buildings and the deputy authority for monument preservation, supported by a thorough expert

Südspitze Kesselhaus mit Stockwerksbau von Hans Müller, Aufnahme 1998
Southern Pinnacle of boiler-house with floor construction by Hans Müller, photograph 1998

Kesselhaus Verfallszustand, Aufnahme 2001
Boiler-house in a state of decay, photograph 2001

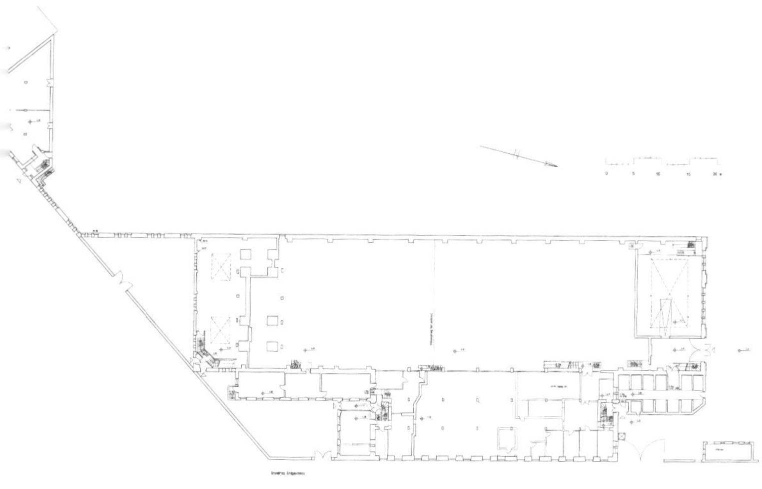

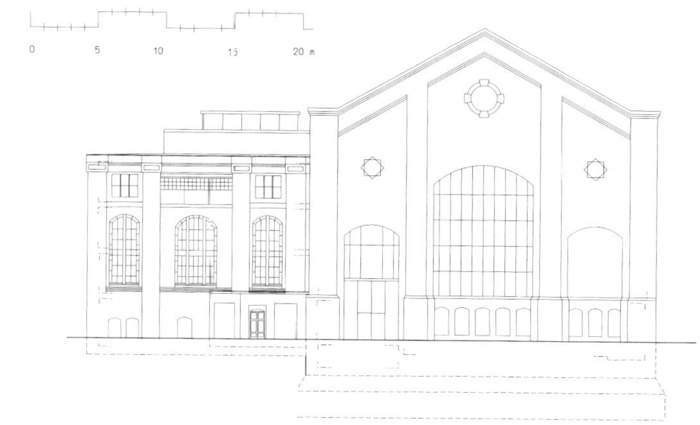

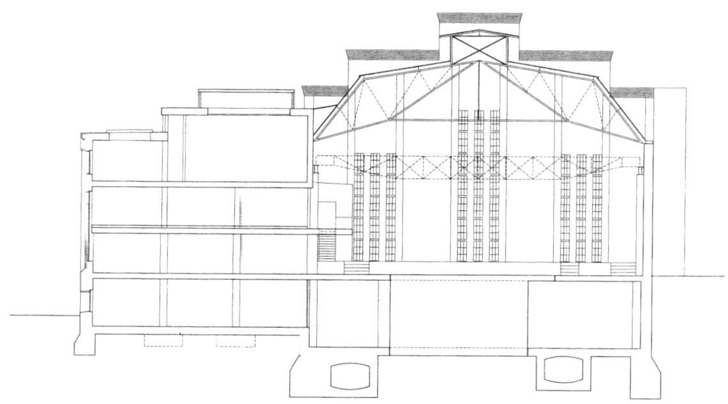

Historisches Ensemble heute, mit Maschinenhalle und Betriebsgebäude:
li. oben: Grundriss, re. oben: Nordansicht, rechts: Querschnitt, aktuelle Zeichnungen, 2003
Historical ensemble today, with machine hall and operations buildings:
upper left: ground plan; upper right: northern view; right: cross-section, contemporary drawings, 2003

Unterer Denkmalschutzbehörde, die durch ein gründliches Bausubstanz- und Instandsetzungsgutachten unterstützt wurden, machten 1997 den Weg frei für eine einvernehmliche Lösung mit einer Teilabrissgenehmigung für das Kesselhaus. Bis August 2001 wurde der baufällige Teil der Halle hinter der Müller'schen Fassade abgetragen.

Das historische Erscheinungsbild des Ensembles, das insbesondere von der Köpenicker Seite den Auftakt für eine eindrucksvolle historische Industrieszene entlang der Köpenicker Chaussee bildet, ist bewahrt. Vollständig und in ihren wesentlichen architektonischen Elementen im ursprünglichen Zustand erhalten geblieben sind die Maschinenhalle mit Betriebsgebäude, das Pförtnerhaus und Teile der Einfriedung. Die Maschinenhalle steht leer. Die historische Wandgestaltung mit eingelassener Ziegelornamentik und große Teile der Fußbodenplatten sind erhalten. Seit 2001 ist das Gelände vollständig beräumt und bietet zusammen mit den denkmalgeschützten Gebäuden einen umfangreichen und entwicklungsfähigen Freiraum für neue Nutzungen.

report on the building's condition and usage, cleared the way in 1997 for a joint solution of a permit for the partial demolition of the boiler-house. The unstable section of the hall behind Müller's façade was dismantled by August 2001.

The historical appearance of the ensemble is preserved. This image especially forms the incentive for an impressive historical design from the Köpenick side of the ensemble, along Köpenick Chaussee. The machine hall, the operations building, the porter's lodge and parts of the fence are intact and largely preserved in the original architectural state. The machine hall now stands empty. The historical structuring of the wall, with the inlaid brick ornamentation and large parts of the floor tiles are preserved. Since 2001, the site is totally cleared and the listed buildings provide an extensive open space for development and new uses.

Historisches Ensemble heute, mit Maschinenhalle und Betriebsgebäude:
Ansicht von der Rummelsburger Landstraße, aktuelle Zeichnung, 2003
Historical ensemble today, with machine hall and operations buildings:
view from Rummelsburger Landstraße, contemporary drawings, 2003

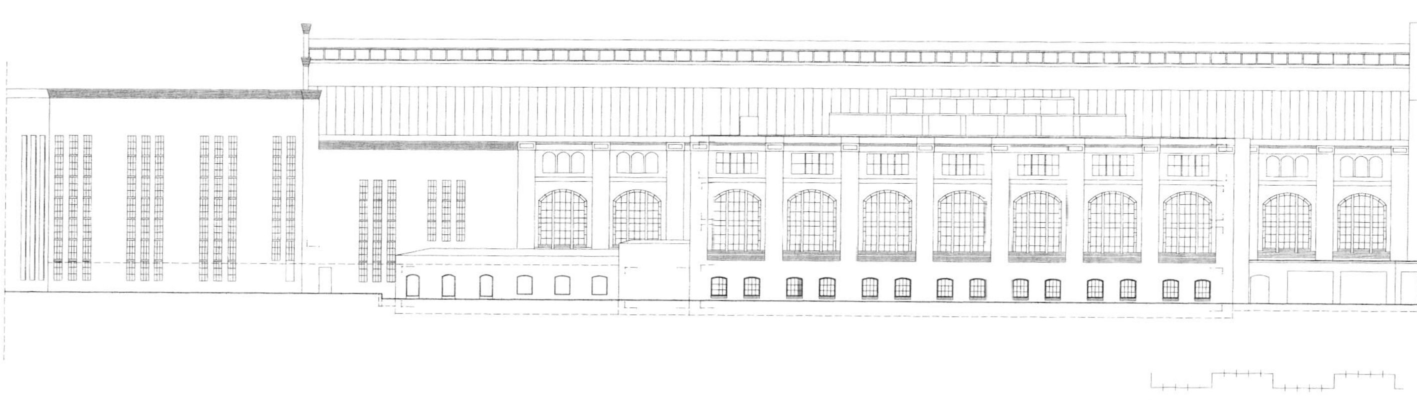

Kraftwerk Rummelsburg • Rummelsburg Power Station, Rummelsburger Landstr. 2-12, 12459 Berlin	
Grundstücksfläche gesamt Land Area, in total	44.073 m²
Vermarktbare Flächen Marketable Sites	
Grundstücksfläche Land Area	44.073 m²
Bruttogeschossfläche (BGF) gesamt Gross Floor Area, in total (BGF)	5.400 m²
mit: Maschinenhalle with: Machine Hall	2.300 m²
Betriebsgebäude Operations Building	3.100 m²

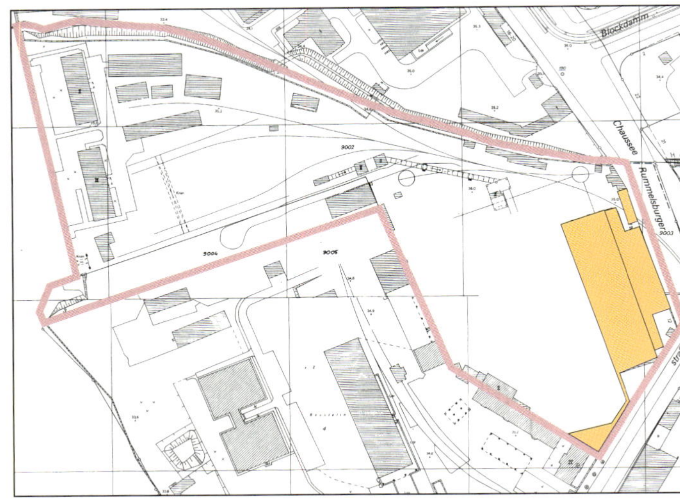

vermarktbare Grundstücksfläche — Marketable Land Area
Denkmalschutz — Listed Building Preservation

Die planungsrechtliche Situation

Planungsrechtlich liegt das Gebiet im Bestimmungsbereich des §34 BGB. Die Neufassung des Flächennutzungsplans von 1998 weist den Standort des alten Kraftwerks Rummelsburg sowie die nördlich und östlich jenseits der Köpenicker Chaussee angrenzenden Flächen als gewerbliche Baufläche aus. Nach Süden sind mit Beginn des ausgedehnten Kleingartengebiets Grünflächen vorgesehen. Entlang der Spree soll ein durchgängiger, gewässerbegleitender Grünstreifen verlaufen. Bei einer angestrebten Entwicklung des Gebiets ist von einer Modifizierung des Flächennutzungsplans auszugehen.

The Legal Planning Situation

In legal planning terms, the area comes under the jurisdiction of § 34 of Federal Law (BGB). The new 1998 edition of the area utilisation plan designates the location of the old Rummelsburg Power Station, as well as the adjacent areas north and west of Köpenicker Chaussee as commercial building-land. To the south-west, green territory is planned from the start of the extended allotment area. Along the Spree, a continual green belt is to run adjacent to the water. It is likely that a modification of the utilisation plan for this area will result from the attempt to develop the location.

Ensemble von Norden,
Rummelsburger Landstraße,
Aufnahme um 1998
Ensemble, view from the north,
Rummelsburger Landstraße,
photograph around 1998

Ensemble von der Köpenicker Seite, Aufnahme 1991
Ensemble, view from Köpenick side, photograph 1991

Städtebauliche Zielsetzung – Nutzungs- und Entwicklungspotenziale

Mit der geplanten Vermarktung des Standorts des ehemaligen Rundfunkgeländes Nalepastraße südlich des Kraftwerksgrundstücks wurde Ende 1999 die Aufstellung eines Bebauungsplans beschlossen. Unter Mitwirkung des Landes Berlin wurde eine städtebauliche Rahmenplanung beauftragt, die auch das Grundstück des alten Kraftwerks Rummelsburg, das direkt am Eingang des Betrachtungsgebiets liegt, mit einschloss. Auf den positiven Standortfaktoren wie der unmittelbaren Wasserlage, den Blickbeziehungen zum Plänterwald, den vielfältig entwicklungsfähigen Flächen und den historischen Gebäuden aufbauend, wurde ein breites Spektrum von Nutzungsszenarien mit den Schwerpunktthemen Freizeitpark, Musik-Entertainment, Campus Hochschule, Logistikzentrum, Fachmarkt-Zentrum und Gewerbe-Wassersport entwickelt. Die Konzepte schienen grundsätzlich umsetzbar, boten aber teilweise keine ausreichende wirtschaftlich tragbare Perspektive. Da keine konkreten Nutzer gebunden werden konnten, wurde das Verfahren 2001 vorläufig zurückgestellt.

Die Ausstrahlung der alten Kraftwerksgebäude bot in der jüngsten Vergangenheit Anlass für verschiedene kulturelle Zwischennutzungen: Filmemacher fühlen sich durch das katakombenähnliche Untergeschoss der Maschinenhalle inspiriert, aber auch für Veranstaltungen aus der Kunst- und Musikszene war das Ensemble von Interesse. Ein greifendes Konzept für temporäre Nutzungen könnte den Grundstein für den langfristigen Erhalt des Industriedenkmals legen und damit sein Überleben sichern.

Für eine zukünftige tragfähige Umnutzungsstrategie wird sich das Grundstück im Wesentlichen auf seine eigenen Qualitäten beziehen müssen, da aus der Umgebung kaum Impulse für eine Initialzündung zu erwarten sind. Das Herauslösen des Grundstücks aus seiner inneren Geschlossenheit kann allerdings nur in einem städtebaulichen Kontext gefunden werden, der das Gebiet an die Stadt und das Umland anbindet und das entstandene Vakuum mit neuen Funktionen wieder belebt. In diesem Prozess wird das alte Kraftwerk mit seiner herausragenden Architektur und seiner städtebaulichen Position als Portal zu dem historischen Industrieraum die neue Identität des Ortes qualitätsvoll bestimmen und damit seine Entwicklung maßgeblich fördern.

Urban Construction Target, Utilisation and Development Potential

At the end of 1999, a construction plan was agreed to coincide with the planned marketing of the Nalepastraße, the site of the former radio station, south of the power station land. In cooperation with the federal state of Berlin, an urban framework plan was commissioned that included the site of the old Rummelsburg Power Station, located right at the entrance of the area under observation. A broad spectrum of usage options was developed due to the site's positive qualities, its direct situation on the water, views towards the Plänter woods, the diversity of the area to be developed, and the historical buildings. Re-utilisation schemes focused on the themes of a leisure park, music and entertainment, a college with a campus site, a logistics centre, a specialised market centre and commercial water sports. Whilst the concepts appeared basically feasible, they partly provided insufficient economic viability. The procedure was temporarily downgraded in 2001, as no definite users could be engaged.

The atmosphere in the old power station buildings offered a recent opportunity for various interim cultural purposes: film-makers feel inspired by the lower-ground level of the machine hall that resembles catacombs; and art and musical events are also interested in this location. A convincing concept for temporary uses could secure the basis for longer-term preservation of the industrial site, thus securing its survival.

The future viability of the re-utilisation strategy will have to be mainly based on the specific qualities of this location, since it can hardly be expected that the initiative and potential will emerge from the surrounding area. However, to extract this land from its inner unity can only be a task for urban construction that connects this area to the city and its surrounding area and injects new life into the present vacuum with new functions. The old power station, with its outstanding architecture and urban position as a gateway to the historical industrial space can define the new identity of this location and significantly promote its own development process.

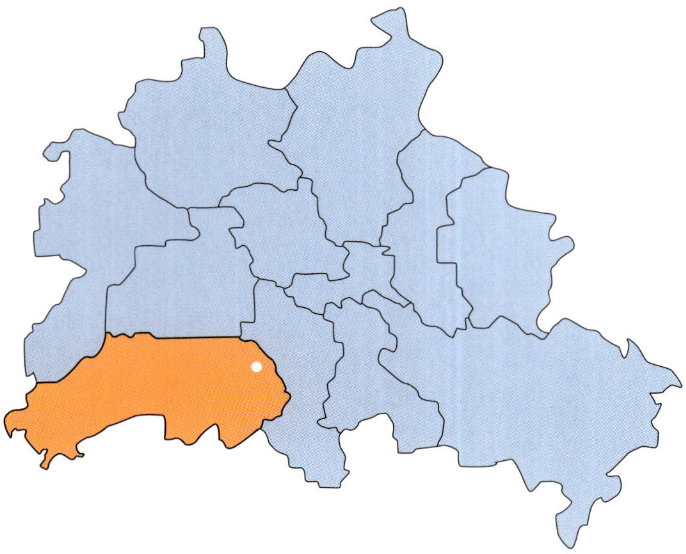

Kraftwerk Steglitz
(Öltanks in den 1990er Jahren
demontiert)
Steglitz Power Station
(oil tanks, dismantled in the 1990s)

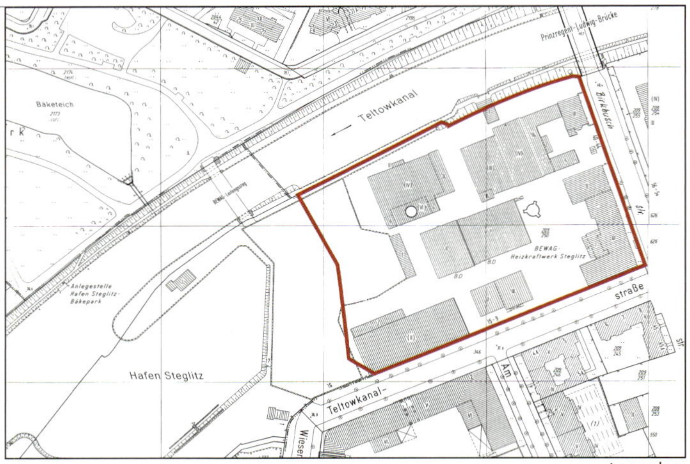

Lageplan
Site plan

DAS KRAFTWERK STEGLITZ
STEGLITZ POWER STATION
Birkbuschstraße 40-44

Die stadträumliche Einbindung

Das Kraftwerk Steglitz markiert mit seiner Lage an der Prinzregent-Ludwig-Brücke den Übergang zwischen den nördlich angrenzenden Wohngebieten und einer kleineren Gewerbezone südlich des Teltowkanals. Von der Brücke und der gegenüberliegen Uferpromenade aus entfaltet das Ensemble mit seinen roten Backsteinfassaden, besonders aber mit seinem hohen Hallengiebel, der den Wasserlauf des Kanals weithin dominiert, eine eindrucksvolle Wirkung.
Das Grundstück ist eingebettet in eine heterogene Stadtstruktur. Die Umgebung ist geprägt durch die südlich an das Gewerbegebiet angrenzende, teilweise villenartige Wohnbebauung von Lankwitz und Lichterfelde Ost, durch die nördlich bis zum Zentrum Steglitz sich verdichtende Wohnbebauung sowie durch einen sich weitläufig entlang des Teltowkanals erstreckenden Grünzug, der über den Stadtpark Steglitz und den Bäke-Park bis zum Schlosspark Lichterfelde reicht. Inmitten dieses Grünzuges liegt in unmittelbarer Nähe zum Kraftwerk das Klinikum Steglitz, daran angrenzend dehnt sich das Villenviertel Lichterfelde-West aus.
Der Standort ist durch den nahe gelegenen Autobahnanschluss sowie eine gute Anbindung an den Nahverkehr günstig erschlossen.

Urban Integration

The location of Steglitz Power Station on Prinzregent-Ludwig bridge defines the transition between the adjacent residential areas to the north and the smaller commercial area south of the Teltow Canal. The ensemble makes a striking impression from the bridge and the opposite riverside promenade, with its red brick façades and especially its high-gabled halls that largely dominate the canal-side area.
The land is embedded in a heterogeneous urban structure. The neighbouring area is defined to the south of the commercial sector by the districts of Lankwitz and Lichterfelde Ost, partly built up with villa-style residences; to the north, as far as the centre of Steglitz, lies a dense residential area as well as a green belt that stretches along most of the Teltow Canal, across Steglitz city park and Bäke Park, as far as the castle park of Lichterfelde. In the middle of this green zone, in the immediate vicinity of the power station is Steglitz Clinic and, nearby, the extensive villa district of Lichterfelde-West.
The location is well served by the nearby connection to the motorway and by good links to local transport.

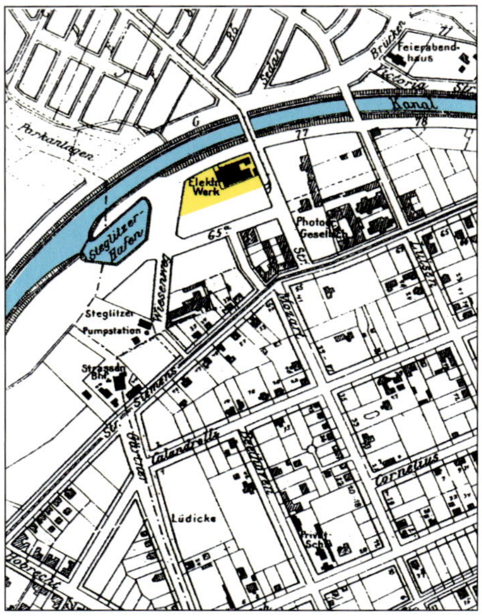

Übersichtsplan 1911 1:15000
Overview plan 1911

1911

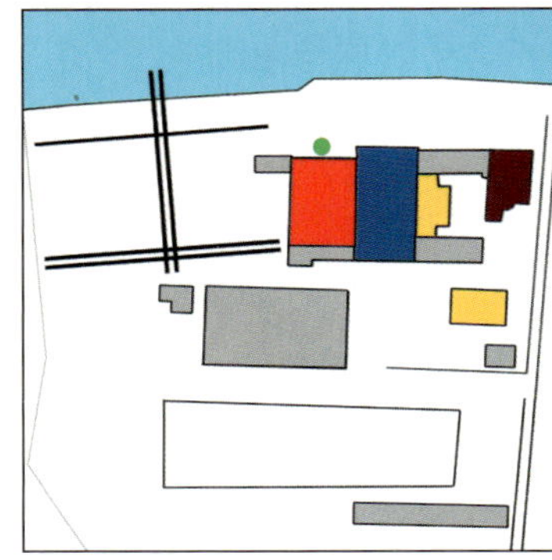

1930

Entwicklungsstufen des Kraftwerks
Power station's development in phases

Wasserstraßen	Waterways
Kesselhaus	Boiler-house
Maschinenhaus	Machine Hall
Stromverteilung	Electricity Distribution
Freiluftanlagen	Open Air Sites
Stromspeicherung	Electricity Storage
Umweltschutz	Environmental Protection
Heizölversorgung	Heating Oil Supply
Wohnen/ Verwaltung	Accommodation/ Administration
Nebengebäude aller Art	Various Outbuildings

Die Entwicklung des Kraftwerks

Die Entstehung eines kommunalen Drehstromkraftwerks in den Jahren 1910/11

Nach umfangreichen Voruntersuchungen beschloss der Gemeinderat Steglitz Ende 1909, ein eigenes kommunales Drehstromkraftwerk zu errichten und gab damit einen entscheidenden Impuls für die weitere Entwicklung der damaligen Vorortgemeinde. Das Kraftwerk wurde in den Jahren 1910/11 auf einem unbebauten Gelände der Gemeinde östlich des Steglitzer Hafens errichtet. Voraussetzung war der Ausbau des Teltowkanals 1906, der zur Kohlenanlieferung und Kühlwasserversorgung benötigt wurde.

Die Planung wurde vom Gemeindebauamt koordiniert. Mit der maschinentechnischen Projektierung des Kraftwerkes wurde der Gemeindeingenieur und spätere Kraftwerksdirektor Martin Rehmer beauftragt, der bauliche Entwurf wurde von Hans Müller in der von ihm seit 1909 ausgeübten Funktion als Gemeindebaumeister erarbeitet.

Beide sollten später in den 1920er Jahren Spitzenstellungen bei der Bewag einnehmen: Mit dem Zusammenschluss der Vorortgemeinden

The Development of the Power Station

The Creation of a Municipal Alternating Current Power Station in 1910/11

After extensive preliminary studies, Steglitz parish council decided, at the end of 1909, to establish its own community alternating current power station and in so doing it provided a decisive incentive for the further development of what was then a suburban parish. The power station was built in 1910/11 on an undeveloped site belonging to the parish to the east of Steglitz Harbour. The condition was the development of Teltow Canal in 1906, as the canal was needed to supply coal and cooling water.

The planning was co-ordinated by the parish's building authority. The commission for machine technology, planned for the power station, was given to the parish engineer and later the director of the power station, Martin Rehmer. The construction design was the work of Hans Müller, who acted as the chief construction engineer in the community from 1909. In the 1920s, both men were later to take leading positions in the Bewag: the integration of the suburban parish into Greater Berlin led in 1923 to Martin Rehmer becoming director of the newly established "Berlin Urban Electricity Works Ltd." (Berliner Städtische Elektrizitätswerke Aktien Gesellschaft). A year later, he employed Hans Müller as the director of the Bewag construction office. Müller was to oversee an extensive building programme for energy distribution within a completely new and extended concept for energy supply. During his time as chief architect at the Bewag, Hans Müller built, between 1924 and 1930, over 50 transformer stations support structures, converter and rectifying works.

Steglitz Power Station came on stream in 1911 with a plant capacity of 6,000 V and an output of 2,500 kW. Shortly before World War One, during 1913/14, the output of the electricity works was increased to 6,000 kW. A steam turbine dating from this period was in service until 1977 and since 1986 the machine has been in the German Technical Museum in Berlin.

Anlage von der Birkbuschstraße aus, Aufnahme um 1911
Site from Birkbuschstraße, photograph around 1911

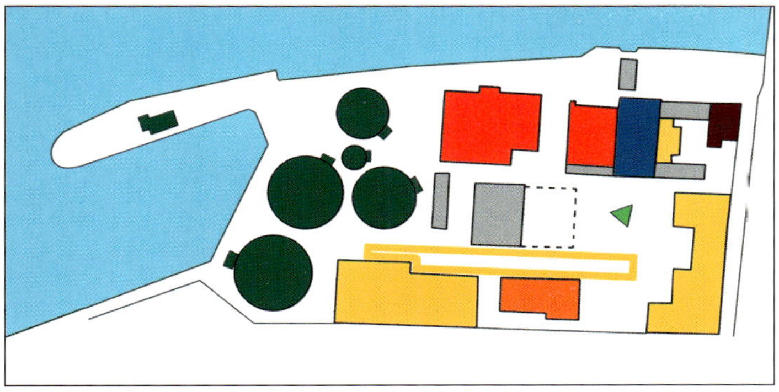

1965

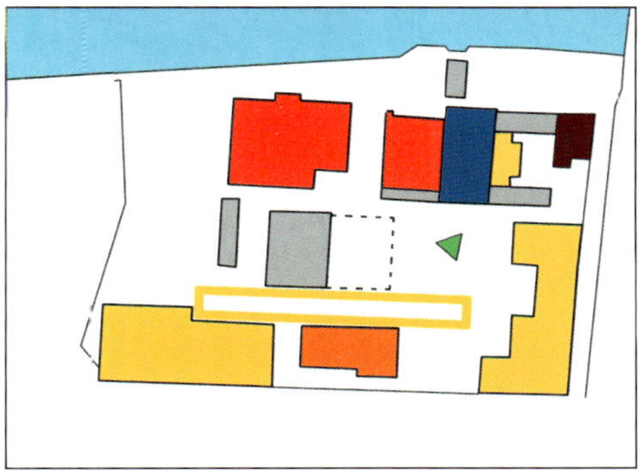

2001

zu Groß-Berlin wurde Martin Rehmer 1923 Vorstand der neu gegründeten „Berliner Städtische Elektrizitätswerke Aktien Gesellschaft". Im folgenden Jahr berief er Hans Müller zum Leiter des Bewag-Baubüros; dieser sollte ein umfangreiches Bauprogramm für die Energieverteilung innerhalb eines völlig neuen erweiterten Energieversorgungskonzeptes realisieren. In seiner Zeit als Chefarchitekt der Bewag baute Hans Müller in den Jahren von 1924 bis 1930 über 50 Abspannwerke, Stützpunkte, Umform- und Gleichrichterwerke.
Das Kraftwerk Steglitz ging 1911 bei einer Betriebsspannung von 6.000 V und einer Leistung von 2.500 kW ans Netz. Kurz vor dem Ersten Weltkrieg, in den Jahren 1913/14, wurde die Leistung des Elektrizitätswerks auf 6.000 kW erhöht. Eine Dampfturbine aus dieser Zeit war bis 1977 in Betrieb und steht seit 1986 im Deutschen Technikmuseum Berlin.

Der Umbau zum Heizwerk und die Erweiterungen in den Jahren 1926 bis 1935

Mit der Bildung von Groß-Berlin erfolgte 1922 der Eigentumsübergang des Kraftwerks an die Stadt Berlin.
In den 1920er Jahren war das Kraftwerk Steglitz Wegbereiter für die Fernwärmeversorgung. So wurde hier erstmals Heißwasser als Wärmeträger eingesetzt, was sich als wesentlich effektiver als Dampf erwies. Das „Städteheizwerk" Steglitz ging 1927 als technische Innovation in Betrieb und wurde bis 1935 kontinuierlich ausgebaut.
Im Zuge der Erhöhung der Verteilungsspannung musste 1929 der Neubau eines 30/6-kV-Abspannwerks errichtet werden. Dieses Gebäude wurde, zusammen mit einem Pförtnerhäuschen neben dem Kraftwerksensemble an der Birkbuschstraße, von Egon Eiermann konzipiert, welcher damals Mitarbeiter der Bewag-Bauabteilung unter Hans Müller war. Die sichtbare Stahlskelettkonstruktion des Kleinabspannwerks war zur Betonung gegen die nicht tragende Prüßwandausfachung schwarz gestrichen. Beide Gebäude gelten als die ersten gebauten Werke Eiermanns.

Das Kraftwerk in und nach dem Zweiten Weltkrieg: Die Jahre von 1940 bis 1959

In den Jahren 1940/41 wurde im Rahmen des kriegsbedingten Ausbaus der Fernstromversorgung der Neubau eines 30/6-kV-Abspannwerks erforderlich. Hierzu musste das Kraftwerksgelände fast um das Doppelte nach Süden hin erweitert werden. Nebengebäude, auch das Pförtnerhaus von Egon Eiermann, wurden abgerissen. Das neue Werk wurde durch das Bewag-Baubüro ausgeführt. Der kolossartige, geschlossene viergeschossige Mauerwerksbau mit Putzfassaden prägt bis heute die Ecke Birkbuschstraße/Teltowkanalstraße.

Reconstruction to a Thermal Plant and Extensions from 1926 to 1935

The creation of Greater Berlin meant that in 1922, ownership of the power station was transferred to the city of Berlin.
Steglitz Power Station was the pioneer in the 1920s for long-distance thermal energy supply. Here, super-heated water was first used as a conductor of heat, since it proved considerably more effective than ordinary steam. The "City Thermal Energy Works" Steglitz went into operation in 1927 as a technical innovation and was continually developed until 1935.
In 1929, in the wake of the increase of distribution capacity, a 30/6 kV installation had to be built from scratch. The building and the porter's lodge, near the power station ensemble at Birkbuschstraße, was designed by Egon Eiermann, who was at that time a member of the Bewag construction department under Hans Müller. The visible steel-skeleton construction of the small plant was painted black in order to emphasise it against the steel rods reinforcing the unsupported wall. Both buildings are considered as Eiermann's first building projects.

The Power Station during and after World War Two: From 1940 to 1959

In 1940/41, the development of thermal heating energy supply, due to the war, meant constructing a new 30/6 kV power plant. For this reason, the power station site had to be extended to almost twice its size towards the south. Outbuildings, as well as Egon Eiermann's porter's lodge, had to be demolished. The new installation was built by the Bewag construction office. The colossal form of the enclosed, four-storey stonework building, with plaster façades, to this day still characterises the corner of Birkbuschstraße/Teltowkanalstraße.

Blick vom Teltowkanal, Aufnahme um 1920
View from Teltow Canal, photograph around 1920

Südansicht vor der Erweiterung, mit dem Straßenbahndepot und dem Abspannwerk von Egon Eiermann, Aufnahme um 1936
Southern view before extension, with tram depot and transformer work by Egon Eiermann, photograph around 1936

Das von Egon Eiermann errichtete Kleinabspannwerk, ein Bau der Neuen Sachlichkeit, wurde in den Neubau integriert, überformt und ist seitdem nicht mehr sichtbar.
Die Reparatur der Kriegsschäden, insbesondere der Wiederaufbau der Dachkonstruktionen, dauerte bis Anfang der 50er Jahre.

Die letzte Erweiterung des alten Kraftwerks in den Jahren von 1959 bis 1963

Mit dem Bau des neuen Gasturbinenspitzenlastkraftwerks 1959 wurden die Anlagen von Kohle auf Ölfeuerung umgestellt. Auf der Fläche des ehemaligen Kohlenlagerplatzes entstand die erste Gasturbinenanlage in Berlin, die zweite im Bundesgebiet.
Südlich des Kesselhauses wurde ein fensterloser kubischer Baukörper mit enormen Ausmaßen errichtet, der die Silhouette der Kraftwerksanlage zum Teltowkanal entscheidend beeinträchtigt. Die Maschinenhalle ist mit gelben Klinkern und hellblauen Keramikplatten verblendet und wird durch einen 80 Meter hohen Schornstein überragt.
Der erweiterte Ausbau der Fernwärmeversorgung, der den Anschluss des Botanischen Gartens und des Universitätsklinikums zum Ziel hatte, führte in den Jahren 1962/63 zur letzten maschinentechnischen Erweiterung des alten Kraftwerks. Im Maschinenhaus wurde eine 25.000-kW-Heizmaschine aufgestellt, die auch einen wesentlichen Umbau im Inneren der Halle mit sich brachte. Im Zuge dieser Erweiterung wurde der alte gemauerte Schornstein zum Teltowkanal abgerissen und durch einen 120 Meter hohen Stahlbetonbau mit drei Blechzügen auf der Hofseite ersetzt. 1964 wurden dann im Zuge der Bevorratung Berlins die drei bereits vorhandenen Öltanks durch zwei weitere ergänzt.

The small installation built by Egon Eiermann in the style of "Neue Sachlichkeit" ("New Objectivity") was integrated into the new building, remodeled and has not been visible since then.
Repairs of war damage, especially the rebuilding of the roof construction, lasted until the beginning of the 1950s.

The Last Extension of the Old Power Station from 1959 to 1963

The building of the new gas turbine peak demand power station in 1959 led to the conversion of installations from coal to oil firing. On the site of the former coal depot, Berlin's first gas turbine plant was built, the second of its kind in the entire country.
To the south of the boiler-house, a windowless, cubic building structure was built on an enormous scale, so that the silhouette of the power station site towards the Teltow Canal was decisively affected. The machine hall is covered with yellow clinker and light-blue ceramic tiles and overshadowed by a 80 metre high chimney.
The extended development of thermal heating energy supply that was meant to join the Botanical Gardens and University Clinic culminated in 1962/63 in the last technical extension of the old power station. In the machine shop, a 25,000 kW heating machine was built that also required considerable modification of the interior of the hall. In the course of this extension, an old stonework chimney towards the Teltow Canal was demolished and replaced, on the courtyard side, by a 120 metre high steel and concrete structure with three metal sheets. In 1964, in the process of reassessing Berlin's energy needs, two additional oil tanks were added to the three already in existence.

Kleinabspannwerk von Egon Eiermann, Aufnahme um 1929
Small transformer work by Egon Eiermann, photograph around 1929

Neubau 30-kV-Abspannwerk, Ansicht Birkbuschstraße, historische Zeichnung
New construction of 30 kV transformer work, view from Birkbuschstraße, historical drawing

Neuer Schornstein,
120 m hoch, im Bau 1962
New chimney,
120 metres in height,
under construction 1962

Die Neubauten nach 1970 und die Stilllegung des Kraftwerks im Jahr 1994

Nach der letzten Erweiterung der 110-kV-Freiluftschaltanlage 1965 entstand in den Jahren 1970/71 der Neubau einer 110/10-kV-Schaltanlage. Der Bau, eine fensterlose Halle in Stahlbetonfertigteilkonstruktion, steht an der südlichen Grundstücksgrenze entlang der Teltowkanalstraße. Mit seiner Inbetriebnahme wurde die alte Freiluftanlage mit Ausnahme der Transformatoren abgerissen.

Im Zuge des Kraftwerks-Modernisierungsprogramms der Bewag in den 1980er Jahren wurde 1985/86 mit dem Bau der Batteriespeicheranlage die letzte Erweiterung am Standort durchgeführt. Dieser Bau war aufgrund der Insel-Lage Berlins notwendig geworden. Die Anlage diente zur Frequenzregulierung und zur Sicherung von Spitzenlasten im Netz. Mit dem Anschluss Berlins an das westdeutsche Verbundnetz 1994 wurde die Anlage dann nicht mehr benötigt. Ihre Außerbetriebnahme beendete die kurze Ära einer weltweiten technischen Innovation, die weder Vorbilder noch Beispiele kannte.

Die verschärften Auflagen des Umweltschutzes hatten seit den 1970er Jahren zu einem kontinuierlichen Rückfahren der Kraftwerksanlagen geführt. Steglitz war nur noch Fernheizwerk und versorgte über die seit 1972 gebaute Wärmeverbundleitung gemeinsam mit dem neuen Heizkraftwerk Lichterfelde den gesamten Süden im Fernwärmenetz. 1994 wurde das Kraftwerk nach 83-jähriger Betriebszeit stillgelegt.

Ansicht vom Teltowkanal vor Errichtung der Gasturbine, Aufnahme um 1959
View from Teltow Canal, before installation of gas turbine, photograph around 1959

The New Buildings from 1970 and the Closure of the Power Station in 1994

After the last extension of the 110 kV open air switching plant in 1965, a new 110/10 kV switching plant was built in 1970/71. The building, a windowless hall in a steel-concrete pre-fabricated structure, is situated on the southern periphery of the site, along Teltowkanalstraße. The operation of the new plant led to the dismantling of the old, open air installation, with the exception of the transformers. The 1980s modernisation programme for the Bewag power stations led in 1985/86 to the construction of a battery accumulator unit that was the last extension to be carried out on this site. Due to Berlin's "island status", it was necessary to build this facility that served frequency regulation and the guarantee of peak demands in the grid. In 1994, Berlin's connection to the west German distribution network meant that the facility was no longer needed. Its closure ended the short era of a worldwide technical innovation that neither had previous models nor other examples.

From the 1970s, the stricter regulations on environmental protection led to the continual downgrading of power station sites. Steglitz only fulfilled the function of a long-distance installation and supplied the entire grid through the 1972 thermal energy transmission network, along with the new Lichterfelde thermal energy power station. In 1994, the power station was closed after 83 years in service.

Gasturbinenanlage vom Teltowkanal aus, Aufnahme um 1963
Gas turbine plant from Teltow Canal, photograph around 1963

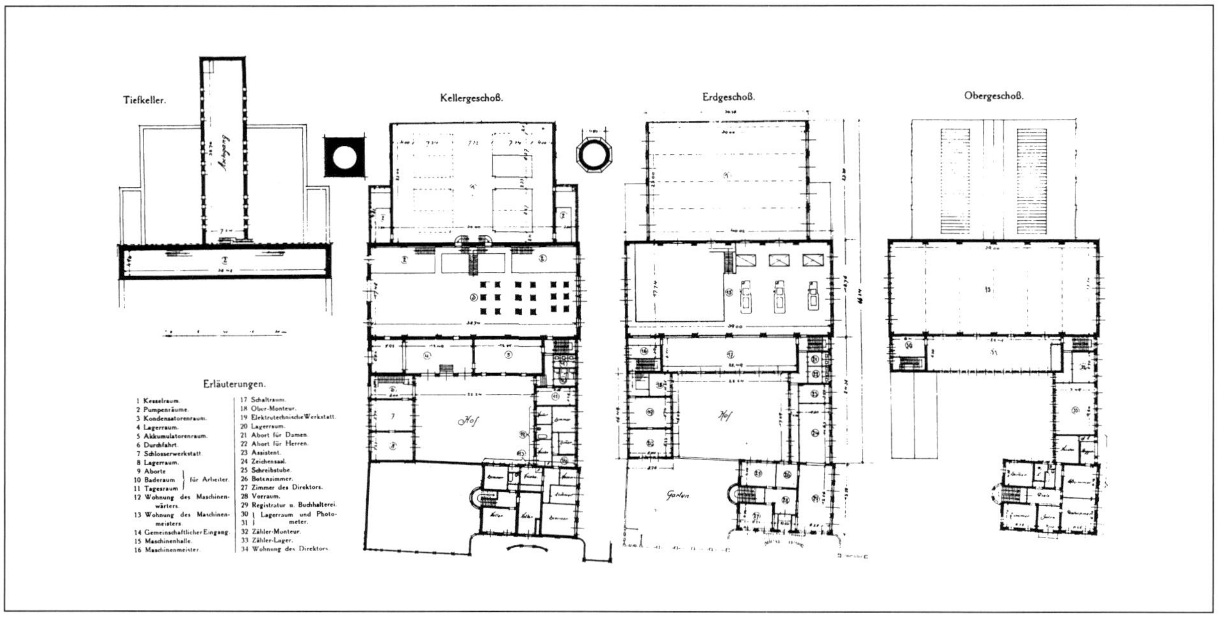

Historische Grundrisse
Historical ground plans

Das Denkmalensemble

Kern der heute noch bestehenden Kraftwerksanlage ist das denkmalgeschützte Ensemble aus den Jahren 1910/11, geplant vom damaligen Gemeindebaumeister und späteren Bewag-Chefarchitekten Hans Müller.
Das Kraftwerk Steglitz gehört der dritten Kraftwerksgeneration in Berlin an, die vor allem von den großen kommunalen Drehstromkraftwerken aus der Zeit unmittelbar vor dem Ersten Weltkrieg repräsentiert wird. In dieser Phase wird der Kraftwerksbau erstmals als ganzheitliche Gestaltungsaufgabe aufgefasst, die über die Fassadengestaltung hinaus vor allem die Anordnung der Baukörper und das Erscheinungsbild im Stadtraum thematisiert.
Das Ensemble ist ein Frühwerk von Hans Müller und gleichzeitig sein erster Industriebau. Es weist mit seiner prägnanten Formensprache schon unverkennbar die Handschrift auf, die der Architekt dann in den 1920er Jahren mit seinen Bauten für die Bewag virtuos entfaltet. Scharf umrissene Baukörper, hohe Lisenengliederungen, Rundbogenfensterreihen und große Thermenfenster mit schmalen, vertikal aufragenden Fensterbändern geben dem Kraftwerk einen sachlich-repräsentativen Charakter. Das Ensemble hat bis heute seine stadtbildprägende Kraft, die durch die Lage am Teltowkanal mit seinen Spiegelungen noch verstärkt wird, nicht verloren. Die bauliche Gliederung entspricht dem Prozess der Umwandlung von Kohle in elektrische Energie: Nach der Anlieferung der Kohle auf den Kohlenhof von der Hafenseite im Westen erfolgte deren Verbrennung im Kesselhaus, die Stromerzeugung im Maschinenhaus und schließlich die Verteilung im angrenzenden Schalthaus, das mit seinen zwei Seitenflügeln einen Lagerhof bildet. Den straßenseitigen Abschluss des Ensembles bildet das Direktorenhaus mit der Wohnung des Kraftwerksdirektors im Obergeschoss, der Registratur und Buchhaltung im Untergeschoss.
Das Maschinenhaus im Zentrum der Anlage liegt quer zum Teltowkanal. Es wird durch seine Höhe und formale Ausgestaltung mit symmetrisch aufgebauten und übergiebelten Schaufronten zur Wasserseite und zum Hof betont. Die Stirnseiten sind durch vertikale Fensterbänder optisch zu Stützen aufgelöst. Dem breiten, durch schmale Lisenen dreigeteilten Mittelfenster entspricht im Giebel eine halbrunde Lünette. Der Giebel ist antikisch mit einem Zahnschnittfries gerahmt. Die schmalen Wandpfeiler zeigen als oberen Abschluss bündig im Mauerwerk liegende geometrische Ornamente aus weißen Ziegelsteinen. Im Inneren ragt die Halle als stützenfreier Raum über zwei Untergeschossen bis auf eine Höhe von 24 Metern empor. Die Längs-

The Listed Building Ensemble

The core of the power station site that still stands today is the listed ensemble dating from 1910/11, planned by Hans Müller, the chief construction engineer of the parish who later became the head architect at the Bewag.
Steglitz Power Station belongs to Berlin's third generation of power plants that are primarily represented by the large municipal alternating current works dating from the period directly before World War One. In this phase, power station construction was seen as a unified task for the first time. The construction work extended beyond the façades and was mainly focused on the organisation of the buildings and urban image.
The ensemble is an early work by Hans Müller that is also his first industrial construction. Its characteristic formal style already indicates the unmistakable signature that the architect developed in virtuoso style in the 1920s with his buildings for the Bewag. Clearly defined buildings, high anchor-block structures, round-arch window series and large-scale thermaly insulated windows with narrow, vertically reaching window tracts all give the power station an objective and imposing character. To this day, the ensemble has not lost its image-defining power that is even enhanced by the urban setting by the situation on the Teltow Canal with the water's reflections. The building structure corresponds to the process of converting coal into electrical energy: after the coal's supply to the coal yard from the harbour side to the west, it was burned in the boiler-house that led to electricity production in the machine shop; and finally, it was distributed in the nearby switching house that forms a storage yard with its two side wings. The completion of the ensemble towards the street forms the director's house with accommodation for the power station's director in the top-floor level and the registrar and accounts office in the lower level.
The machine shop in the centre of the site lies at an angle to the Teltow Canal. It is emphasised by its height and formal equipment with symmetrically built and gabled main façades on the waterfront and towards the courtyard. The end walls are optically tapered to supports by vertical window tracts. The broad middle window, divided into three by narrow anchor blocks corresponds to the semi-circular lunette window in the gable. The gable is framed in antique style by a tooth-cut frieze. As the completed upper section, the narrow wall pillars show closely placed

Ansicht Maschinenhalle
vom Hof, Aufnahme 1989
View of machine hall
with courtyard, photograph 1989

seiten sind durch einfache vorgeblendete Wandpfeiler in Rundbogensegmente gegliedert, die ehemals die Schienen des Laufkrans trugen Darüber, unterhalb des Dachs ist eine schmale Fensterreihe angeordnet. Konstruktiv und künstlerisch ausgereift sind die filigrane Stahlbinderkonstruktion, die feinen, unten als Rechteck, oben im Rautenmuster gestalteten Stahlfensterteilungen als auch die Lünette mit radialen Stegen.

Das Kesselhaus, eine niedrige, breit gelagerte Halle mit Lichtbändern im flachem Satteldach, tritt mit seiner bescheidenen formalen Gestaltung deutlich hinter dem repräsentativen Aufbau des Maschinenhauses zurück. Die Längsfassaden sind mit breiten Fenstern und hinterlegten Lisenen gegliedert. Als sichtbares Stahlfachwerk mit einfacher Ausmauerung ist der nach Westen orientierte Giebel belassen. Der südliche, zweigeschossige Anbau stammt aus der Erweiterung von 1928.

Zwischen Maschinenhalle und Wohnhaus vermittelt ein dreigeschossiger, längsgestreckter Bürotrakt, dessen horizontaler Zug durch eine Blendarkade, die Erdgeschoss und erstes Obergeschoss zusammenfasst, noch unterstützt wird. Zum Stillstand kommt diese Bewegung im abschließenden Direktorenhaus, einem ausgewogenen, symmetrisch aufgebauten kubischen Baukörper mit Walmdach, zentralem Eingangsvorbau und einem seitlich ausschwingenden, halbrunden Treppenhaus. Die Fassade des dreigeschossigen Hauses wird durch eine große Pilasterordnung auf Sockeln gegliedert.

Der Gebäudekomplex, der durchgehend in rotem Backstein errichtet und mit Schiefer – ursprünglich Eternitschiefer – gedeckt ist, weist deutliche Anklänge an eine klassizierende Formensprache und Kenntnis von Karl Friedrich Schinkels Werk auf. Beides wird auch an den späteren Bauten von Hans Müller immer wieder sichtbar. Der sachlich-repräsentative Charakter der Maschinenhalle wird vom Architekten geschickt mit der wohnlichen Wirkung von Verwaltungs- und Direktorengebäude verbunden. Die weißen Ziegelreihen im Ortgang- und Traufbereich und die weißen Ziegelintarsien an der hohen Hallenwand sind eine gestalterische Rarität in Berlin. Das Ensemble überzeugt mit einem in den Proportionen straff durchkomponierten, die unterschiedlichen Funktionen deutlich heraushebenden und dennoch durch die architektonischen Details einheitlich wirkenden Entwurf.

geometrical ornaments in the stonework, made out of white stone bricks. In the interior, the hall extends as a free-standing space over two lower floors, up to a height of 24 metres. The longer sides consist of simple projecting wall pillars in round-arch segments, formerly carrying the tracks for the crane. Above this, underneath the roof, a narrow row of windows is set. The filigree steel-girder construction is productively and artistically completed as are the well-positioned steel window partitions, below as a right-angle and above as angular patterns, as well as the lunette window with radial vertical plates.

The boiler-house, – a broad-based hall with light tracts in the flat sadle roof –, clearly recedes with its formal, modest structure behind the imposing construction of the machine shop. The longer façades consist of wide windows and concealed anchor blocks. The western gable remains as visible steel latticework with simple stonework. The southern two-storey extension dates from the 1928.

A three-storey, extending office wing connects the machine hall and the living quarters. The horizontal character of the wing is still supported by a blind arcade that links the ground floor to the first-floor upper level. This movement is halted in the concluding director's house, a well-balanced, symmetrical and cubic building with a hipped roof, central entrance-hall lobby and a semi-circular stairwell, leading off to the side. The façade of the three-storey house is structured by a large group of pilasters on a pedestal.

The building complex, entirely of redbrick and slates, originally fibre-cement slates, clearly shows tones of a classicist formal style and knowledge of Karl Friedrich Schinkel's work. Both aspects are frequently visible on later works by Hans Müller. The architect skilfully combines the sober representational character of the machine hall with the residential image of the administrative and director's offices. The white rows of tiles in the passageway and eaves area and the white tiles on the high hall-wall are a structural rarity for Berlin. The ensemble is convincing because of tightly regulated proportions throughout that clearly emphasise the various functions and still retain a unified design effect by means of special architectural details.

Ansicht vom Teltowkanal, historische Zeichnung
View from Teltow Canal, historical drawing

Zustand und Nutzung im Jahr 2003

Obwohl das Kraftwerk bis in die 1960er Jahre mehrfach modernisiert und erweitert wurde, blieb die bauliche Hülle der alten Kraftwerksanlage von Hans Müller weitgehend unverändert erhalten.
Das Dach der Maschinenhalle wurde nach dem Zweiten Weltkrieg vereinfacht, ohne die ursprünglichen flachen Fledermausgauben wieder aufgebaut. Der Helm des Treppenhauses zwischen Maschinenhalle und Werkstattflügel wurde nicht wieder errichtet; der Treppenturm schließt heute mit einem flachen Zeltdach ab. Im Zuge einer anstehenden Dachsanierung sollen die Fledermausgauben sowie der Turmhelm originalgetreu wieder aufgebaut werden. Der Innenraum der Maschinenhalle zeigt mit der verlängerten Kranbahn und dem Unterbau für die 25.000-kW-Heizmaschine die Veränderungen aus der letzten technischen Erweiterung von 1962/63 zur Hofseite. Das Direktoren-

Condition and Usage in 2003

Although in the 1960s, the power station was modernised several times and extended, the building shell of Hans Müller's old power station plant was largely unchanged.
The roof of the machine hall was simplified after World War Two and rebuilt without the original, flat dormer windows. The spire of the stairwell between the machine hall and the workshop-wing was not rebuilt; and the stairway tower today has a flat, tent roof. Plans to renovate the roof are to reconstruct the dormer windows and the tower spire. The interior of the machine hall shows – with its extended crane-way and the structure for the 25,000 kW heating machine – those changes made in the last technical extension of 1962/63 towards the courtyard. The

Südansicht mit Kesselhausanbau von 1928, Aufnahme 2002
Southern view with 1928 boiler-house extension, photograph 2002

Blick von Süden über d e Freiluftanlage, Nachkriegsaufnahme von 1955
View from the south over the open-air plant, post-war photograph in 1955

haus ist in seiner Grundrissgestaltung, der Innenausbau mit originalen Türen noch weitgehend erhalten.
Seit der Außerbetriebnahme des Kraftwerkes 1994 sind Maschinen- und Kesselhaus ohne feste Nutzung. In der komplett beräumten Maschinenhalle finden temporär Filmarbeiten statt, im Kesselhaus ist noch die technische Ausrüstung der 1960er Jahre vorhanden. Die Flächen im Bürotrakt und die Wohnungen sind derzeit größtenteils vermietet. Die vom technischen Betrieb befreite Grundstücksfläche weist neben dem Denkmalensemble auch einen erheblichen nicht geschützten Gebäudebestand auf. Dessen Erhalt wird im Rahmen einer Neukonzeption des Standortes abzuwägen sein.

director's house is largely preserved in its original state and the interior has the original doors.
Since the closure of the power station in 1994, the machine and boiler-house have not been regularly used. In the machine hall that was completely cleared, film projects occasionally take place. In the boiler-house, the technical equipment of the 1960s is still available. The land released by the end of technical operations shows, along with the listed property ensemble, a range of buildings that are not protected as listed buildings. The conservation of these buildings must be evaluated within the context of a new concept for the location.

Maschinenhalle: Untergeschoss, Aufnahme 2002
Machine hall: lower ground floor, photograph 2002

Maschinenhalle: Aufnahme 2002
Machine hall: photograph 2002

Kraftwerk Steglitz • Steglitz Power Station, Birkbuschstr. 40-44, 12167 Berlin	
Grundstücksfläche gesamt Land Area, in total	24.079 m²
Vermarktbare Flächen Marketable Sites	
Grundstücksfläche Land Area	14.500 m²
Bruttogeschossfläche (BGF) gesamt Gross Floor Area, in total (BGF)	11.260 m²
darunter: Denkmalensemble including: Listed Building Ensemble	4.850 m²
mit: Maschinenhalle with: Machine Hall	1.500 m²
Kesselhaus Boiler-house	700 m²

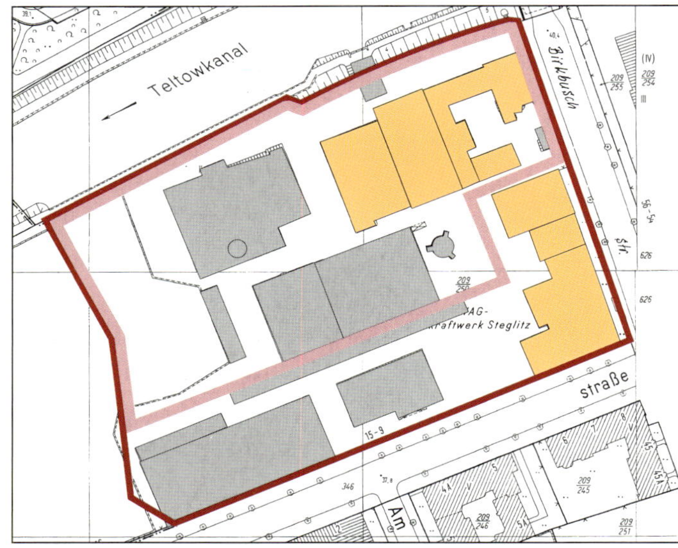

vermarktbare Grundstücksfläche — Marketable Land Area

Denkmalschutz — Listed Building Preservation

Die planungsrechtliche Situation

Für das Grundstück des alten Kraftwerks Steglitz sowie für das jenseits der Birkbuschstraße liegende Areal zwischen Teltowkanal und Siemensstraße gilt die Baunutzungsverordnung von 1960 in Verbindung mit der BauOBln 1958 mit der Ausweisung: „Beschränktes Arbeitsgebiet, Baustufe III/3“. Das südlich angrenzende kleine Gewerbegebiet liegt im Geltungsbereich von rechtskräftigen Bebauungsplänen. Die Grundstücke zwischen Birkbuschstraße, Teltowkanalstraße, Wiesenweg und Siemensstraße weisen eine Mischnutzung aus Wohnen und Gewerbe auf. Unter anderem haben sich hier auf dem Gelände der ehemaligen Berthold AG, einem denkmalgeschützten Altbau, in den letzten Jahren Unternehmen der Medizintechnik und angrenzender Technologiefelder angesiedelt. Das Areal weist zur Siemensstraße hin eine Erweiterungsfläche auf. In den Gewerbegebieten am südlichen Teltowkanalufer, jenseits der Birkbuschstraße und südlich des Hafens Steglitz liegen große Flächen brach oder werden nur extensiv genutzt. Im Zusammenhang mit der anstehenden Entwicklung des Gebietes werden die Festlegungen der verbindlichen Bauleitplanung zu überprüfen sein.

The Legal Planning Situation

The 1960 building utilisation order and the 1958 “BauOBln” designates “Limited Commercial area, building level III/3”. This directive applies to the site of the old Steglitz Power Station, as well as for the area beyond Birkbuschstraße between the Teltow Canal and Siemensstraße. The adjacent small commercial zone to the south comes under the jurisdiction of legally binding construction plans. The land between Birkbuschstraße, Teltowkanalstraße, Wiesenweg and Siemensstraße is used for a mixture of residential and commercial purposes. Amongst other things, in recent years, on the site of the former company, Berthold AG and a listed old property, medical technology and related companies have located their business. The area towards Siemensstraße shows expansion potential. In the commercial areas on the south bank of Teltow Canal, beyond Birkbuschstraße and south of Steglitz Harbour, large areas lie empty or are extensively used. In connection with the planned development of the area, directives on legal site management are to be examined.

Kraftwerk Steglitz, Ansicht vom Teltowkanal, Aufnahme 2002
Steglitz Power Station, view from Teltow Canal, photograph 2002

Nutzungs- und Entwicklungspotenziale

Mit der Stilllegung des Kraftwerksbetriebes eröffnen sich für den Standort Steglitz völlig neue Entwicklungsoptionen. Im Herzen eines sich im Umbruch befindenden Gewerbegebietes gelegen, könnte er, insbesondere mit einer Öffnung zum Hafen Steglitz, zum Initialzünder für einen Strukturwandel auch der benachbarten Brachflächen werden. Die im Jahr 2002 erschienene Veröffentlichung der Bewag „Elektropolis – Chancen und Visionen" präsentiert Umnutzungsstudien von Studenten der Technischen Fachhochschule Berlin zum Abspannwerk Scharnhorst und zum Kraftwerk Steglitz. Die vorgestellten Entwürfe zum alten Kraftwerk zeigen ein breites Spektrum virtueller Nachnutzungsvarianten aus den Bereichen Kunst/Kultur, Freizeit/Sport, Gastronomie/Hotel, Bildung und Begegnung sowie einen möglichen Hochschulcampus. Alle Entwürfe werden als grundsätzlich umsetzbar und genehmigungsfähig eingestuft. Indem sie beim Betrachter über die Denkmalgegenwart hinausgehende Zukunftsvisionen mobilisieren, könnten sie Basis und Ansatzpunkte für potenzielle Investoren liefern.

Das alte denkmalgeschützte Kraftwerksensemble Steglitz trägt mit seiner prägnanten Silhouette zur Prinzregent-Ludwig-Brücke und zum Teltowkanal wesentlich und qualitätvoll zur Identität des Ortes und damit für seine zukünftige Entwicklung bei.

Utilisation and Development Potential

When the power station was closed down, completely new development options were available for the Steglitz site. Situated in the heart of a changing commercial area, the location could, especially due to its access to Steglitz Harbour, become the catalyst for structural change in the neighbouring waste land.

The Bewag publication that appeared in 2002, "Electropolis – Opportunities and Visions" presents the reutilisation studies by students of Berlin's Technical College for the plant at Scharnhorst and for Steglitz Power Station. The promoted designs for the old power station suggest a broad range of virtual reutilisation schemes from the areas of art/culture, leisure/sport, hotel/catering, education and getting together as well as a possible college campus. All designs are basically feasible and classified as valid for authorisation. The inspiration they provide to the viewer of visions for the future that go beyond the present condition of the listed properties could supply the basis and incentives for potential investors.

The old listed status of the Steglitz power station ensemble contributes – in a vital and qualitative way, with its distinctive silhouette on the Prinzregent-Ludwig bridge and towards the Teltow Canal – to the identity of this location and therefore it promotes its own future development.

Kraftwerk Charlottenburg, Maschinenhalle, Galerie mit Schaltwarte, Aufnahme um 1900
Charlottenburg Power Station, machine hall, gallery with switching observation post, photograph around 1900

SPIELRÄUME FÜR EINE DENKMALERHALTENDE UMNUTZUNG
OPTIONS FOR RE-USE AND PRESERVATION

Matthias Dunger

Das Kraftwerk Charlottenburg

Mit der endgültigen Stilllegung im Jahr 2001 steht die komplette Anlage eines klassischen Großkraftwerks mit Kesselhaus, Maschinenhaus und Nebengebäuden aus verschiedenen Entwicklungsphasen des Kraftwerksbaus zur Disposition. Die gesamte Ausstattung, beginnend mit den ältesten Turbinen von 1929 für die Ruths-Speicheranlage über die Technik aus der Zeit der Modernisierung der 50er und 60er Jahre bis zur modernen Rauchgasentschwefelungsanlage von 1989, ist noch vorhanden. Wie in allen Kraftwerken ist auch hier die Maschinenhalle als Herzstück der Anlage am besten erhalten. Zuletzt wurde sie 1929 für die Turbinen der aus konservatorischer und technikgeschichtlicher Sicht bedeutsamen Ruths-Speicher verlängert. Die repräsentative Schaufassade, die äußere als auch die innere Wandgliederung der Längsseiten und die elegant geschwungenen Dachbinder bilden die bewahrenswerten Qualitäten, die in einem künftigen Nutzungskonzept zur Geltung kommen sollten. Hervorzuheben ist auch die architektonische Auflösung der Giebelwand für die ehemalige Schaltwarte. In den monumentalen Kesselhäusern wäre nach dem Entfernen der Ausstattung im gesamten Volumen ein Ausbau vorstellbar, im Rest des alten Kesselhauses sollte jedoch die alte Dachkonstruktion berücksichtigt werden. Während für das Kesselhaus der 50er Jahre die Fassadengliederung mit den Belichtungsöffnungen und teilweise auch den Geschossebenen vorgegeben sind, bedarf der Kesselhausneubau der 60er Jahre für diese Zwecke größerer architektonischer Überlegungen und Eingriffe. Es ist deshalb durchaus zu bedenken, ob nicht dieses große Raumvolumen, auch über eine konventionelle Verwendung hinaus, als Ganzes genutzt werden kann. Für das architektonisch besonders bemerkenswerte und anspruchsvolle 30-kV-Schalthaus ist unter Berücksichtigung der Haupterschließungen und Treppenhäuser ein Ausbau mit neuen Grundrissen möglich und wünschenswert. Die besonders charakteristischen Stahlfenster sind aber auch hier zu bewahren.

Charlottenburg Power Station

After the definitive closure in 2001, the entire installation of a classical high-capacity power station, complete with boiler-house, machine-house and various outbuildings from different phases of the power station's development are now available. All the equipment – from the 1929 oldest turbines for the Ruth accumulator-unit, through to modernised 1950s and 1960s technology, to the latest technology for removing sulphur from flue-fumes installed in 1989 – is all on-site. As in all power stations here, too, the machine shop, the centrepiece of the whole installation, is best preserved. This was last extended in 1929 for the turbines of the Ruth accumulator-unit – important for reasons of conservation and technological history. The imposing façade, both the external and internal wall-division of the long sides, and the stylishly undulating roof-trusses constitute the most enduring qualities that should come to the fore in any future re-utilisation. Also noteworthy is the architectonic breaking up of the gable-wall for the former switching station. In the monumental boiler-houses, after removing all the equipment throughout the building, an extension would certainly be possible; though in the rest of the old boiler-house the existing roof-construction needs to be respected. Whereas in the case of the 1950s boiler-house the divisions in the façade, which shows the presence of light-slits and some indication of different storey-levels, must be taken as given, the 1960s boiler-house reconstruction will need a more radical approach to architectural ideas and interventions. Accordingly, one should seriously consider whether this enormous space cannot be utilised as a whole – if need be, in a non-conventional form. For the architecturally very striking and ambitious 30 kV switching station, an extension following new ground plans, as long as it respected the main developments and stairwells, would be both possible and desirable. Even in this case, however, the particularly characteristic steel windows should be retained.

Kraftwerk Rummelsburg, Erweiterungsbau von H. Müller, Aufnahme von 1998
Rummelsburg Power Station, extension by H. Müller, photograph in 1998

Das Kraftwerk Moabit

Der bereits zwischen den Jahren 1952 und 1958 technisch völlig erneuerte Teil II des Moabiter Kraftwerks mit Maschinenhalle, Kesselhaus und dem ab 1929 hinzugefügten Schalthaus wurde 1992 endgültig stillgelegt. Sowohl die Turbinensätze als auch die technische Ausstattung des Kesselhauses sind noch vorhanden, aus konservatorischer Sicht wird eine Erhaltung jedoch nicht gefordert. Während im breit gelagerten Kesselhaus bei Erhaltung der Fassade und der historischen Dachkonstruktion eine räumliche Unterteilung möglich erscheint, sollte auch in der Maschinenhalle eine anderweitige Nutzung den ursprüngliche Raumeindruck mit der großzügigen Giebelverglasung zur Geltung bringen. Mit Ausnahme der originalen Dachbinder sind im Inneren keine bemerkenswerten historischen Details erhalten. Das Schalthaus von Klingenberg und Issel ist einschließlich der Schaltzellen komplett beräumt und kann einem anderweitigen Nutzungskonzept entsprechend im Inneren umgestaltet werden.

Das Kraftwerk Rummelsburg

Bemerkenswert und noch immer eindrucksvoll ist die Maschinenhalle des Kraftwerks mit den noch vorhandenen Resten der technischen Ausstattung. Dazu gehören die Kranbahn mit beiden Kranbrücken und ganz besonders die Schaltbühne mit der Kulisse der Regeltechnik. Alle Details, die die architektonische Sorgfalt und den gestalterischen Aufwand für diesen Raum illustrieren, die originalen Wandoberflächen, der Fliesenboden oder die Reste der Geländereinfassungen, müssen in geeigneter Weise überliefert werden. Auch hier sollte die Maschinenhalle in ihrer Gänze erlebbar bleiben, was nicht ausschließt, dass im begrenzten Umfang Einbauten von Galerien, eingeschossigen Pavillons oder dergleichen vorgenommen werden könnten. Im seitlichen Verwaltungsgebäude ist unter Bewahrung der Treppenhäuser und der noch erhaltenen historischen Türen eine Neuordnung der Räum-

Moabit Power Station

Section II of Moabit Power Station, which had technically been completely renovated, and now boasted a machine shop, boiler-house and – from 1929 onward – an additional switching station, was closed down completely in 1992. Both the sets of turbines and all the technical equipment in the boiler-house are still in existence, but no conservation measures will be needed for these items. Whilst it is quite possible to envisage dividing up the broad space of the boiler-house, keeping the façade and the historic roof-construction, in the machine shop, too, utilising the space in a different way may well bring out the full effect of the original space with its generous gable-windows. Apart from the original roof-trusses no noteworthy historic details still exist on the inside. The switching station, designed by Klingenberg and Issel, has a full complement of rooms, including the "switching cells", and can be reconceived appropriately on the inside as part of a different utilisation on a broader scale.

Rummelsburg Power Station

What is most notable and impressive here is the power station's machine shop, with the existing remains of the technical equipment. This includes the crane-way, both crane-bridges and, especially, the "switching stage" with the regulatory mechanisms alongside. All the details that illustrate the architectural scrupulousness and the creative effort that have gone into this space – i.e. the original wall-surfaces, the tiled floor, the vestiges of the banister borders – must be retained in an appropriate manner. Here, too, the machine shop should be experienced as a whole, which does not preclude the limited construction of galleries, one-storey pavilions, or such like. In the administrative building to one side, provided that the stairwells and the remaining historical doors are retained, a restructuring of all the rooms could then be under-

Kraftwerk Rummelsburg, Maschinenhalle, Aufnahme von 1998
Rummelsburg Power Station, machine hall, photograph in 1998

lichkeiten im Interesse einer besseren Nutzbarkeit möglich. Besonderer Wert muss auf die Erhaltung der bauzeitlichen Fenster gelegt werden, die entscheidend zur gestalterischen Identität des Gebäudes beitragen.

Durch den Abbruch des Kesselhauses ist eine weitläufige Fläche entstanden, die für eine zusätzliche Bebauung bereitsteht. Dabei ist ganz besonders auf die sinnvolle Einbindung der vom Architekten Hans Müller errichteten Fassade der Erweiterungsbauten des Maschinen- und Kesselhauses einschließlich des dreieckigen Werkstattgebäudes der späten 20er Jahre zu achten. Die als Sichtfassade gestaltete Außenseite des Maschinenhauses könnte dabei auf geeignete Weise zur Geltung gebracht werden. Die Neubebauung sollte die Kubatur des ehemaligen Kesselhauses in Höhe und Länge nicht überschreiten.

Das Kraftwerk Steglitz

Nur vier Jahre nach Baubeginn des von den Berliner Elektrizitätswerken konzipierten Kraftwerks Rummelsburg hatte der Architekt Hans Müller für das Steglitzer Kraftwerk eine ganz andere, am Neoklassizismus der Zeit orientierte Architektursprache gefunden. Die denkmalpflegerische Bedeutung der Anlage liegt deshalb auch wesentlich in ihrer architektonischen Qualität und erfordert eine besondere Rücksicht auf das äußere Erscheinungsbild. Anbauten sowie zusätzliche Wandöffnungen, die die Silhouette beziehungsweise das sonst ungestörte Erscheinungsbild wesentlich berühren, sollten unbedingt vermieden werden.

Herzstück der Anlage ist die Maschinenhalle, die mit der endgültigen Stilllegung des Kraftwerks 1994 komplett beräumt wurde. Jede wie auch immer geartete Nutzung sollte sich die Raumqualitäten der Halle zu Eigen machen. Eine Veränderung durch das Einziehen von Decken oder Trennwänden ist hier kaum denkbar, eventuell ist noch das Einstellen eines kleineren selbständigen Baukörpers oder begrenzter Galerieebenen möglich, die jedoch immer das Raumerlebnis weitestgehend ungestört lassen.

Im Gegensatz dazu bleibt für die Halle des Kesselhauses die Option offen, nach Entfernen der technischen Ausstattung, deren Bewertung aus denkmalpflegerischer Sicht eine Erhaltung und Konservierung nicht rechtfertigt, funktionelle Einbauten vorzunehmen. Die Giebelwand des Kesselhauses aus schlichtem, mit Klinkern ausgemauertem Stahlfachwerk deutet darauf hin, dass die Möglichkeit der Verlängerung

taken in the context of using the space to greater advantage. Particular attention must be paid to retaining the windows, contemporary to the original building, which materially help establish its identity.

The demolition of the boiler-house has left an extensive site clear for an additional construction. In designing this, it is especially important effectively to include the façade, constructed by the architect Hans Müller, of the extensions to the machine hall and boiler-house and not to forget the triangular workshop building, erected in the late 1920s. The outside of the machine shop, constructed as a "show-façade", could effectively – in an appropriate style – form part of any subsequent redesign. Any new constructions should not exceed the cubature of the former boiler-house in height or length.

Steglitz Power Station

Only four years after construction started on Rummelsburg Power Station, conceived by the Berliner Elektrizitätswerke, the architect Hans Müller found a totally different architectural style for Steglitz Power Station – a style rooted in then contemporary neoclassicism. Accordingly, the importance of the installation as a listed building lies primarily in its architectonic quality, demanding that particular attention be paid to its external appearance. Any additional structures or wall-openings that radically alter the silhouette or the building's otherwise undisturbed external appearance should be avoided at all costs.

The installation's centrepiece is the machine hall, which was completely cleared in 1994, when the power station was closed down. Any future utilisation – in whatever form – should embody this machine shop's spatial qualities. Here, for example, one can hardly conceive of adding ceilings or partition walls; though the possibility still exists of adding a small, self-contained extra building or limited gallery-levels, as long as both leave the spatial experience more or less intact.

By contrast, as far as the hall for the boiler-house is concerned, the option remains open, – after removing the technical equipment, whose retention and conservation are not justified in terms of preserving monuments –, to undertake functional extra construction. The boiler-house gable wall, made of simple steel latticework, faced with clinker, indicates that an extension should not be ruled out. If need be, there would

Kraftwerk Steglitz mit Giebel des Maschinenhauses, Aufnahme von 1978
Steglitz Power Station with the machine hall's gable, photograph in 1978

offen gehalten werden sollte. Bei Erfordernis bestünde hier ein größerer Spielraum in der Öffnungsstruktur der Fassade, im Gegensatz zu den übrigen, bereits auf typische Hans Müller'sche Art und Weise "backsteinscharf" durchgestalteten Fassaden, die verändernde Eingriffe kaum vertragen.

Die Grundrisse von Wohn- und Verwaltungsgebäude sind unproblematisch. Hier können begrenzte Veränderungen im Inneren wie beispielsweise das Entfernen oder Verschieben nicht tragender Wände, das Verlegen von Türöffnungen sowie weitere Veränderungen konzeptabhängig vorgenommen werden, sofern die Haupterschließungen, Treppenhäuser und Eingangsbereiche mit ihren bauzeitlichen Details erhalten bleiben. Größere Bewegungsfreiheiten ergeben sich auch im Bereich der Trafozellen, deren Tore großflächig geöffnet werden können. Eventuell kann auch der gesamte Anbau zur Disposition gestellt werden. Eine komplette Hofüberbauung ist jedoch abzulehnen. Der 1959 begonnene Bau des Gasturbinenkraftwerks gehört nicht zum Denkmalbestand, seine Einbeziehung in ein Nutzungskonzept liegt jedoch durchaus im Interesse einer Bewahrung der geschichtlichen Tradition des Standorts. Diese Flächen stehen sonst einer Erweiterung der Nutzfläche oder einer angepassten zusätzlichen Bebauung zur Verfügung.

be more room for manoeuvre here to adapt the façade's structure, which is more "open-ended" than the other façades; these are fully "bricked over" in characteristic Hans Müller style, and they would not allow radical alteration.

The ground plans of the accommodation and administrative buildings present no problems. Here, it is possible to carry out limited, appropriate alterations to the internal fabric, – as, for instance, the demolition or pushing back of non-load-bearing inner walls, the relocation of door-openings, as well as other alterations –, provided that the main developments, stairwells and entrance areas, remain, along with all their original details. There is also more room for manoeuvre in the area of the transformer-cells, whose gates can be widely extended. All things being equal, the whole extension building can be made available. That being said, covering the entire courtyard is not to be encouraged. Although the gas-turbine power plant, begun in 1959, is not among the monuments to be preserved, its inclusion in a re-utilisation plan would greatly help to preserve the whole site's historical traditions. Otherwise, these areas remain fully available for an enlargement of the functional surface or for suitable extra construction.

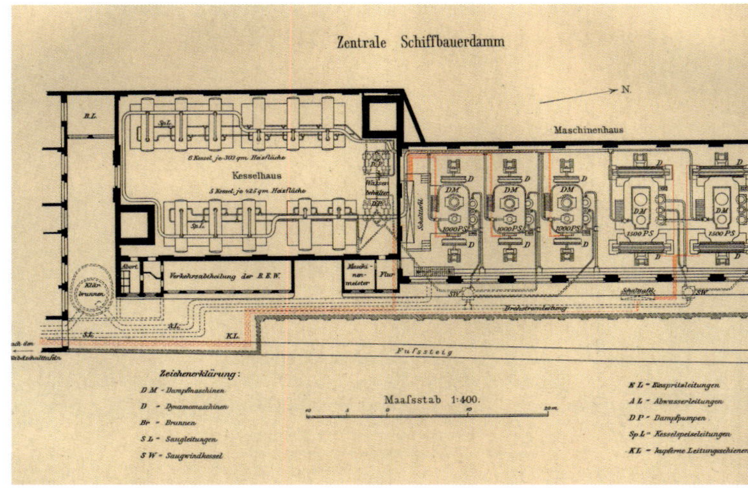

Zentrale Schiffbauerdamm
Headquarters Schiffbauerdamm

DENKMALERHALTUNG DURCH UMNUTZUNG
PRESERVING MONUMENTS BY FINDING NEW USES

Das Kraftwerk Schiffbauerdamm
Schiffbauerdamm Power Station

Hans Achim Grube

Das Gelände des ehemaligen Kraftwerks Schiffbauerdamm war das erste Kraftwerksareal, für das die Bewag eine mögliche Umnutzung geplant und das sie auf dieser Basis verkauft hat. Das Kraftwerk Schiffbauerdamm war zwischen 1889 und 1890 als letzte von vier Stadtzentralen errichtet worden.
Da hier ausreichend Liegenschaftsfläche verfügbar war, wurde das Kesselhaus neben dem Maschinenhaus errichtet. Das Maschinenhaus hatte eine lichte Raumhöhe von 18 Metern, die sich über eine Fläche von 16,8 x 26,5 Metern erstreckte. Das Kesselhaus mit einer Grundrissfläche von 35 x 21 Metern und einer Höhe von 11,5 Metern erlaubte die Anlage der Kessel in Doppelreihe mit einem Schornstein. Am Kesselhaus wurden Anbauten für die Maschinenmeister und weitere Betriebseinheiten der damaligen BEW errichtet. An der repräsentativen Straßenfront am Schiffbauerdamm 22 wurde als Blockrandbebauung das Verwaltungsgebäude der BEW und der AEG errichtet. Auch beherbergte dieses Gebäude, das das Architekturbüro Cremer und

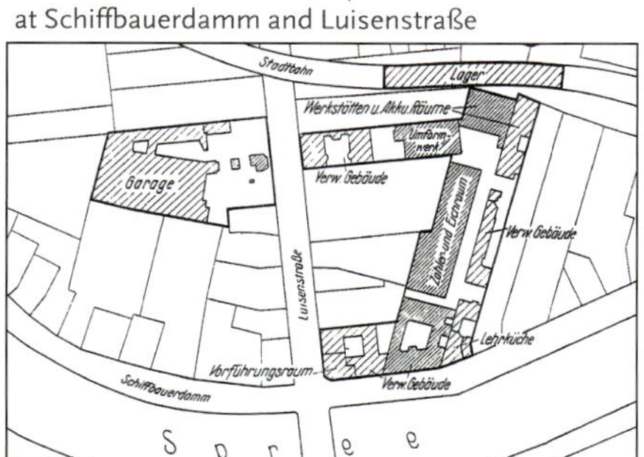

Lage der Hauptverwaltung
am Schiffbauerdamm und Luisenstraße
Site of administrative headquarters
at Schiffbauerdamm and Luisenstraße

The site occupied by the erstwhile Schiffbauerdamm Power Station was the first power-plant land for which the Bewag projected re-utilisation and which it sold with this in mind. Built between 1889 and 1890, Schiffbauerdamm Power Station was the last of four to be constructed as "City HQ" plants.
As there was sufficient land available, the boiler-house was built next to the machine shop. The machine shop boasted a clear height of 18 metres and extended over an area of 16.8 x 26.5 metres. With an area of 35 x 21 metres and a height of 11.5 metres the boiler-house offered enough space for both boilers to be installed in a double row, together with a chimney. Beside the boiler-house other buildings were constructed to accommodate the machine-operators and other mechanics working for what was then the BEW (Berliner Elektrizitätswerke). The administrative buildings of both the BEW and the AEG (Allgemeine Elektrizitäts-Gesellschaft) were built as blocks of offices along the imposing street façade of 22 Schiffbauerdamm. This building, which was designed by the architectural firm of Cremer and Wolffenstein, also contained the directors' official residences which, following the tradition of those days, were very close to the site of professional activity. In addition to office staff, the administrative building at 35 Luisenstraße housed a show room for new electrical household appliances. In the courtyard area, the length of the machine shop, a trial railway line ran, which helped research into electricity-driven vehicles. In April 1926, the power station operation was closed down and subsequently the site was extended for administrative purposes. That being said, the administrative buildings on Schiffbauerdamm were destroyed during World War Two. After the division of Berlin into eastern and western sectors, the East Berlin energy supply authority established itself in the complex of buildings between Schiffbauerdamm and 35 Luisenstraße and was accommodated successively in those buildings that survived. The reunification of Germany meant that the former electricity supplier of

links und rechts: Probeausführung einer elektrischen Hochbahn auf dem Grundstück Schiffbauerdamm-Luisenstraße, Aufnahmen um 1899
To the left and right: Trial run of an electrified elevated railway on the Schiffbauerdamm/Luisenstraße site, photographed in 1899

Wolffenstein entworfen hatte, die Dienstwohnungen der Direktoren, die in damaliger Unternehmertradition direkt an das gewerbliche Firmenareal angrenzten. Das Verwaltungsgebäude Luisenstraße 35 beherbergte zusätzlich zu den Büroeinheiten einen Präsentationsraum für neue Haushaltsgeräte, die mit Strom betrieben wurden. Auf dem Hofgelände gab es entlang des Maschinenhauses eine Versuchsbahn, die der Erforschung von strombetriebenen Fahrzeugen diente. Im April 1926 wurde der Kraftwerksbetrieb eingestellt, anschließend erfolgte der Ausbau des Standortes für Verwaltungszwecke. Die Verwaltungsgebäude am Schiffbauerdamm wurden im Zweiten Weltkrieg jedoch zerstört. Nach der Teilung Berlins in die Ost- und West-Sektoren nahm die Hauptverwaltung des Ost-Berliner Energieversorgungsunternehmens in dem Liegenschaftskomplex zwischen Schiffbauerdamm und Luisenstraße 35 ihren Betrieb auf und wurde sukzessive in den bestehenden Gebäuden etabliert. Durch die Wiedervereinigung Deutschlands wurden die Ost-Berliner EBAG und die West-Berliner Bewag wieder zusammengeführt. Als provisorische Hauptverwaltung wurde bis zum Bezug des neuen Sitzes in Berlin Treptow die Stauffenbergstraße genutzt, das EBAG-Gelände am Schiffbauerdamm jedoch weiterhin für Verwaltungszwecke und die Unterbringung von Werkstätten.

Die Ausgangssituation zur Umnutzung

In einer ersten Nutzungsanalyse für verschiedene Standorte der Bewag wurde Ende 1994 für den Standort Luisenstraße/Schiffbauerdamm ein erhebliches Entwicklungspotenzial aufgezeigt. Die Kernaussage der Studie bestand in dem Nachweis, dass aufgrund der günstigen Lage des Standortes in Berlin-Mitte, insbesondere auch durch die Ansiedlung von Bundestags- und Regierungsfunktionen begründete Zukunftsperspektiven bestehen würden. Die Studie ging davon aus, dass die Attraktivität des Bereichs durch die zahlreichen kulturellen Einrichtungen und die vorhandenen Theater gesteigert würde. Eine Nutzungsanalyse hob außerdem die gute Anbindung an das Nahverkehrsystem über den Bahnhof Friedrichstraße hervor, der in nur geringer Entfernung liegt. Über diesen zentralen Knotenpunkt des BVG-Systems bestehen U- und S-Bahn Verbindungen in alle Bezirke Berlins. Im Flächennutzungsplan wurde das Gebiet als Sonderbaufläche mit Zweckbestimmung „Hauptstadtfunktion“ ausgewiesen.

East Berlin Energieversorgung Berlin AG (EBAG) and West Berlin's Bewag were also recombined. However, until moving into the new headquarters in Berlin-Treptow, they went on using the Stauffenbergstraße premises; with the EBAG offices on Schiffbauerdamm being set aside for adminstrative purposes and as a location for workshops.

Re-utilisation: The Starting Point

In late 1994, an early analysis for various Bewag sites showed considerable development potential for the Luisenstraße/Schiffbauerdamm site. The basic finding of this study was that the favourable site in the city centre (Berlin-Mitte), together with the imminent relocation to Berlin of Bundestag and other government bodies all gave grounds for future optimism. The study presupposed that the attractiveness of the site would be greatly enhanced by the wealth of cultural outlets on offer in Berlin, as well as numerous theatres. An analysis of "re-utilisation potential" also emphasised the superb travel connections for short-hop journeys via Friedrichstraße station, which is on the doorstep. By way of this important junction in the BVG (Berliner Verkehrsgesellschaft) network, the underground and S-bahn transport links run to all districts in Berlin. In the land utilisation plan the whole area was seen as ripe for special development, with a particular emphasis on its "role as a capital city".

Verwaltungsgebäude am Schiffbauerdamm, Zustand um 1900
Administrative building Schiffbauerdamm, condition around 1900

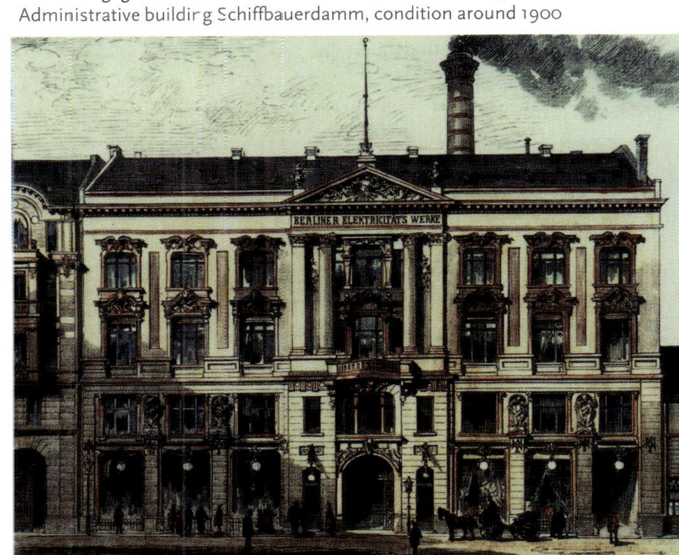

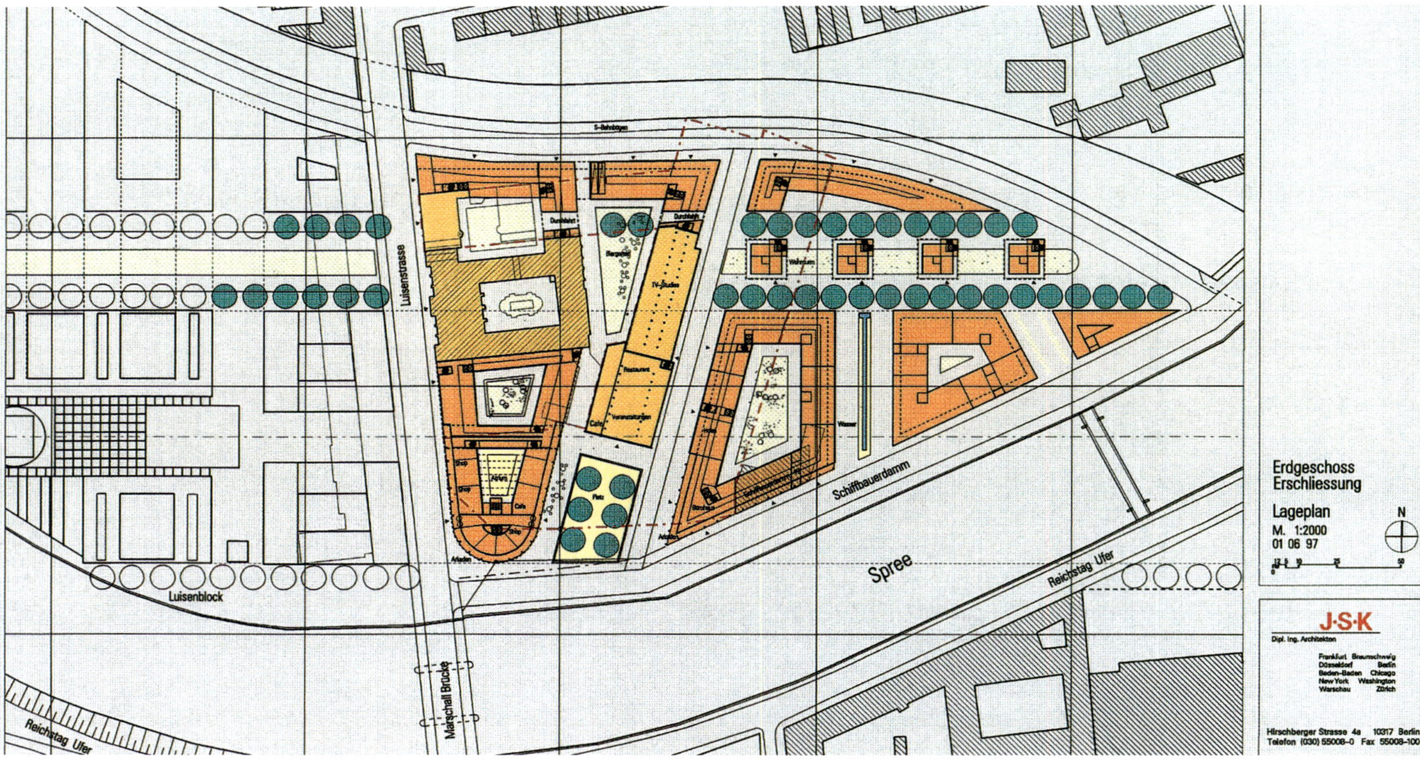

Schiffbauerdamm/Luisenstraße, Lageplan
Schiffbauerdamm/Luisenstraße, site plan

Das Planungskonzept zur Blockentwicklung des Bereiches Schiffbauerdamm/Luisenstraße

Anfang 1997 beauftragte die Bewag das Architekturbüro J·S·K mit einer Projektstudie für die Liegenschaft Schiffbauerdamm/Luisenstraße. Die Planungsstudie sollte sich mit der Entwicklung von Blocktypologien als Abstimmungsgrundlage für Gespräche mit dem Bund, dem Land Berlin und entsprechenden Nutzern beschäftigen und stellte die historische Situation, Lage und städtebauliche Einbindung, Bestand, aktuelle Nutzung, die Erschließung des Quartiers und die Realisierungsstufen der Liegenschaft dar. J·S·K schlug eine städtebauliche Gesamtentwicklung für den Bereich südlich der S-Bahntrasse vor, begrenzt von der S-Bahntrasse, der im Süden verlaufenden Spree und der im Westen liegenden Luisenstraße. Das Büro entwickelte dazu die historische Berliner Blockstruktur weiter, die durch einzelne Solitärbauten ergänzt werden sollte. Der städtebauliche Entwurf erstreckte sich auf einer Fläche von annähernd 30.000 Quadratmetern, wovon lediglich 13.800 auf Bewag-Grundstücke entfielen. Ansonsten waren überwiegend bundeseigene Grundstücke betroffen. Bei einer angenommenen Geschossflächenzahl (GFZ) von 4,0 bis 4,5 wurde eine oberirdische Bruttogeschossfläche (BGF) von annähernd 62.000 Quadratmetern auf den Bewag-eigenen Grundstücken geplant.

Der Abstimmungsprozess über Umsetzungsmöglichkeiten

Basierend auf der Projektstudie für das Quartier Schiffbauerdamm/Luisenstraße des Architekturbüros J·S·K strebte die Bewag Mitte 1997 an, die Realisierungschancen des Projektes mit dem Bund und dem Land Berlin zeitnah abzustimmen. Grundsätzlich hatte die Bewag entschieden, dieses Projekt nicht eigenständig durchzuführen, sondern mit einem institutionellen, finanzstarken Partner. Basierend auf den Vorplanungen von J·S·K wurden mit mehreren potenziellen Investoren Sondierungsgespräche geführt. Im Sommer 1997 trat die Bewag in direkte Verhandlungen mit der für die Baumaßnahmen des Deutschen Bundestages zuständigen Bundesbaugesellschaft. Darüber hinaus entstanden zeitnah Kontakte zwischen der Bewag und dem Fernseh- und Kommunikationsunternehmen RTL, das auf der Suche nach einem Standort für ein Hauptstadtstudio in der Nähe des Reichstagsgebäu-

The Planning Concept for Block-development of Schiffbauerdamm/Luisenstraße

In early 1997, the Bewag commissioned the architectural firm J·S·K to conduct a study-project for the Schiffbauerdamm/Luisenstraße site. The study was meant to achieve several goals. These included developing the technology of blocks as the basis of agreement between the federal government (Bund), the federal state (Land) of Berlin and those potential users who were involved. To do this, it would establish the historical situation, the present position and what involvement the city would have; and it would also examine what buildings remained, their present role, and what steps needed to be taken to realise the project for this site. The J·S·K office suggested an overall urban rebuilding programme for the area south of the S-bahn route, as bounded by the demarcation of the S-bahn route, the river Spree flowing in the south and Luisenstraße situated in the west. The architects further developed the block-structure, traditional to Berlin, which was to be enlarged by a few individual buildings. The outline plan for urban regeneration covered an area of some 30,000 square metres, of which only 13,800 square metres were on land owned by the Bewag. Most of the rest was federally-owned land. Out of an assumed flor space index of between 4 and 4.5, plans were drawn up for a gross flor space of nearly 62,000 square metres on building land owned by the Bewag.

The Procedure for Voting on Possible Re-utilisation

Basing itself on the study-plan for the district of Schiffbauerdamm/Luisenstraße prepared by the J·S·K firm of architects, in mid-1997, the Bewag sought an early clarification of the plan's chances of success from the federal authorities and the Land of Berlin. In principle, the Bewag had already decided not to carry out this project single-handedly but with an institutional, financially strong partner. Basing themselves on the J·S·K projected designs, the Bewag held talks to sound out several potential investors. In summer 1997, the Bewag started direct negotiations with the Federal Building Authority, which was responsible for all building arrangements made by the German Bundestag.

Hauptverwaltung Schiffbauerdamm/Luisenstraße,
links im Hintergrund das 1928 erbaute Bürohaus
Administrative headquarters Schiffbauerdamm/Luisenstraße,
on the left in the background, the office building from 1928

Verwaltungsgebäude Schiffbauerdamm/Luisenstraße,
Aufnahme um 1947
Administrative building Schiffbauerdamm/
Luisenstraße, photograph around 1947

des war. Die Bundesbaugesellschaft erklärte, dass der Bundestag grundsätzlich großes Interesse an der von der Bewag vorgelegten Projektentwicklung habe. Dabei sollte die Gesamtentwicklung in drei Stufen, beginnend mit der Entwicklung der im Bewag-Eigentum befindlichen Grundstücke und mit einer zügigen Realisierung der Stellplatzanzahl, die vom Bund gewünscht war, erfolgen. Die Bewag strebte an, die in ihrem Eigentum stehenden Grundstücksflächen in eine gemeinsame Projektentwicklungsgesellschaft einzubringen, während die IVG die zur Arrondierung notwendigen Grundstücksflächen der Oberfinanzdirektion sowie der Deutschen Bahn AG erwerben und ebenfalls in die Projektentwicklungsgesellschaft einbringen sollte. Während dieser Verhandlungen konkretisierte sich das Interesse von RTL, Teile der denkmalgeschützten Fläche des ehemaligen Kesselhauses und des Werkstattgebäudes als Studio- und Redaktionsräume anzumieten. Um eine erste Initialzündung für das Projekt zu bewirken, entschied die Bewag Anfang 1998, das denkmalgeschützte Kesselhaus und das denkmalgeschützte Werkstattgebäude entsprechend der Planungskonzeption zu sanieren und langfristig an RTL zu vermieten. Diesbezüglich wurde von einem zehnjährigen Mietvertrag mit einer weiteren zehnjährigen Anmietungsoption ausgegangen.

Der Verkauf der Liegenschaft

Mitte 1998 änderte sich die Gesprächs- und Verhandlungsführung des Bundes dahin gehend, dass nunmehr ein Ankauf der Bewag-eigenen Liegenschaften favorisiert wurde. Für die Verkaufsverhandlungen war die Lage der Liegenschaft im Entwicklungsgebiet ausschlaggebend. Ein „freihändiges" Verhandeln des Kaufpreises unter Berücksichtigung der herausragenden Lage in der Nähe des zukünftigen Sitzes des Deutschen Bundestages und weiterer Bundesministerien war für das Entwicklungsgebiet nicht möglich. Erschwerend kam bei der Bewag-Liegenschaft hinzu, dass die Bestandsgebäude im Wesentlichen unter Denkmalschutz standen und der städtebauliche Grundriss problematisch war, weil umfangreiche Verkehrsflächen auf der Liegenschaft langfristig angeordnet werden mussten. Mit dem Kaufvertrag zwischen der Bewag und dem Bund verließ die Bewag einen historischen Betriebsstandort, der seit der Einführung der Elektrizität in Berlin von ihr genutzt worden war.

Additionally, timely contacts were made with the TV and Communication Network RTL, which was looking for a location for a studio in the capital near the Reichstag building. The Federal Building Authority declared that the Bundestag was, in principle, very interested in the urban redevelopment project proposed by the Bewag. According to this, the total development was due to take place in three phases, starting with the sites owned by the Bewag, then following with a brisk redevelopment of all the locations specified by the federal government. The Bewag strove to incorporate all the building land in its possession into a joint-ventur to develop the project, whilst the IVG was to involve the Federal Ministry of Finance as well as German Railways (Deutsche Bahn AG) in re-utilising relevant sites, and also including them in the joint venture. During these negotiations RTL firmed up its interest in renting some sections of the listed erstwhile boiler-house and workshop room as a studio and editorial suite. To kickstart the project, in early 1998, the Bewag decided to agree to renovate, then rent to RTL on a long-term basis, both listed buildings: the boiler-house and the workshop under the terms of the projected plan. Accordingly, the Bewag agreed initially to a ten-year rental contract, with an option to renew for another ten years.

Selling the Site

In mid-1998, the negotiating stance of the federation changed, insofar as purchasing the Bewag-owned sites was now favoured. For the ensuing sales' negotiations the site's location in the Berlin Redevelopment Area was of crucial importance. An "open-ended" negotiation of the purchase price in the light of its superb position near the future home of the German Bundestag and other federal ministries was not possible for the Redevelopment Area. Further complications arose: the existing buildings were mostly listed, and the outline plan for urban regeneration needed to have extensive long-term transport links on-site. In concluding the sale-contract between the Bewag and the federal authorities the Bewag was saying goodbye to a historic operational site which had been used ever since the Bewag had first supplied electricity to Berlin.

Ehemaliges Kraftwerk Schiffbauerdamm, Historische Ziegelwände des Kesselhauses
Former Schiffbauerdamm Power Station, historical brick walls of the boiler-house

Zentrale Schiffbauerdamm, 1. Drehstromanlage, 1896
Headquarters Schiffbauerdamm, first alternating current plant, 1896

Ehemaliges Kraftwerk Schiffbauerdamm, RTL-News-Center Multifunktionsstudio, Aufnahme von 2000
The former Schiffbauerdamm power station, RTL News Center Multi-function Studio, photographed in 2000

Der Umbau des ehemaligen Kesselhauses und des Werkstattgebäudes

Nach dem Verkauf der Liegenschaft kam es, basierend auf den Planungen und Verhandlungen zwischen Bewag und RTL, zu einem Vertragsabschluss zwischen dem Bund und RTL zum Umbau und zur Nutzung des Kesselhauses und des Werkstattgebäudes als Hauptstadtstudio des Fernsehsenders. Nach den noch von der Bewag durchgeführten Abstimmungen bezog sich der Denkmalschutz vor allem auf die Gebäudehülle und die zum Teil im historischen Zustand erhaltenen Fassadenbereiche. Die konstruktive Struktur sollte im Wesentlichen erhalten bleiben. Als neue Nutzung für das ehemalige Kesselhaus war ein zweigeschossiger Büro- und Studiobereich mit Galerieebene geplant. Im ehemaligen Werkstattgebäude sollten im Erdgeschoss technische Räume für den Studiobetrieb eingerichtet werden, in den übrigen drei Etagen waren Büroflächen vorgesehen. Zusätzlich wurde ein etwa 200 Quadratmeter großes Archiv im Kellerbereich des Werkstattgebäudes geplant. Zur Umnutzung für Studio- und Büroflächen mussten beide Gebäudeteile bis auf die tragenden Wand- und Deckenkonstruktionen vollständig entkernt werden. Des Weiteren war ein Abriss des Anbaus an der Vorderseite des Kesselhauses vorgesehen. Die erhaltene Backsteinfassade sollte entsprechend den historischen Ansichten saniert werden. Die vorhandene Dachkonstruktion des Kesselhauses musste bis auf die stählerne Tragkonstruktion vollständig erneuert werden. Darüber hinaus erfolgte der Einbau einer Wärmedämmung entsprechend der Wärmeschutzverordnung. Die alten Fenster in beiden Gebäudeteilen wurden fachgerecht entfernt und durch neue Fenster mit den erforderlichen Schall- und Wärmeschutzwerten ersetzt. Für alle Fenster wurden manuell bedienbare Sonnenschutzrollos vorgesehen. Dabei fanden die Aspekte des Denkmalschutzes und der besonderen Architektur des Objektes Berücksichtigung. Das vorhandene Oberlicht des Kesselhauses ersetzte man durch ein neues Oberlicht in Stahlglaskonstruktion. Zur Vermeidung von direkter Sonneneinstrahlung auf die darunter befindlichen Bereiche und unter Berücksichtigung der für TV-Produktionen notwendigen Lichtverhältnisse wurde getöntes, beschichtetes Glas als Füllung vorgesehen, das zusätzlich eine hohe Schallschutzdämmung gewährleistet. Zur möglichen Verdunkelung baute man elektrisch

The Rebuilding of the Former Boiler-house and the Workshop Building

After the sale of the site, a contract based on the projects and negotiations between the Bewag and RTL was agreed between the federation and RTL to convert and re-utilise the boiler-house and workshop as the television station's capital-city studio. In accordance with the arrangements agreed by the Bewag, the "special architectural interest" listing applied principally to the buildings' shell and façade surfaces, which were to some extent historic. In essence, the building structure was to remain as it was. The conversion planned for the erstwhile boiler-house was a two-storey building, housing offices and a studio, with a gallery-level. As for the former workshop, technical rooms related to the working of the studios were envisaged for the ground floor, whilst the remaining three floors were to be set aside for offices. In addition, an archive, some 200 square metres in size, was proposed for the basement area of the workshop building. In order to accommodate studio and office facilities both sections of the building had to be completely gutted except for the retaining walls and the roof-constructions. Furthermore, a plan was afoot to demolish the extension built on to the front side of the boiler-house. The brickwork façade, which had been retained, was to be renovated on historic principles. The existing roof-construction of the boiler-house needed to be totally renewed down to the steel support-skeleton. In addition, to satisfy the insulation regulations, layers of thermal insulation had to be added. The old windows in both sections of the building were professionally removed and replaced with new windows that met all the regulations about sound-proofing and heat-insulation. For all the windows, manually adjustable sunblinds were envisaged. All in all, then, both the specific aspects of monument-preservation and the special architecture of these particular buildings were respected. The existing overhead light in the boiler-house was replaced with a new celing-light made of steel and glass. To avoid the presence of direct sunlight on the areas immediately underneath, and taking into account the lighting conditions applicable to TV production, tinted, coated glass was given priority – it also made for

Ehemaliges Kraftwerk Schiffbauerdamm, RTL-News-Center Eingangshalle, Aufnahme von 2000
The forner Schiffbauerdamm power station, RTL News Center Entrance Foyer, photographed in 2000

bedienbare Rollos und Blenden. Der Boden im Studiobereich des Kesselhauses wurde als Doppelboden mit Parkettbelag ausgestattet. Für den vorgelagerten Haupteingangsbereich, der gleichzeitig als Wartezone dienen sollte, war ein Bodenbelag in Naturstein vorgesehen. In den verbleibenden Bereichen für Büros, Technikräume, Flure verlegte man Teppichbelag. Der Haupteingangsbereich im Kesselhaus wurde mit einem Windfang als Schleuse und mit einer Stahlglaskonstruktion als Abtrennung zum Studiogroßraum ausgeführt. Die Beheizung der Gebäude erfolgt mittels der bereits auf der Liegenschaft vorhandenen Fernwärmeversorgung der Bewag. Die Baumaßnahmen für RTL wurden innerhalb kürzester Zeit nach dem Verkauf der Liegenschaft geplant und mit ihrer Umsetzung wurde direkt nach dem Vertragsabschluss über die mietweise Nutzung der Fläche begonnen. Die anderen Bestandsgebäude, insbesondere das ebenfalls unter Denkmalschutz stehende historische Verwaltungsgebäude der Bewag im Gebäude Luisenstraße 35, wurden ebenfalls kurz nach dem Abschluss des Kaufvertrages den Notwendigkeiten der Büroräume des Deutschen Bundestages angepasst. Das Vorderhaus war aufgrund der langjährigen Nutzung durch die Bewag nach der Wiedervereinigung in einem bautechnisch sehr guten Erhaltungszustand, sodass hier die baulichen Anpassungen auf die spezifische gebäudetechnische Ausstattung für den Bundestag zu begrenzen waren.

Würdigung der Sanierungskonzeption

Die Sanierungskonzeption und deren Umsetzung gingen sensibel auf die historischen Bestandteile des Industriedenkmals des ehemaligen Kesselhauses ein. Die vom Schiffbauerdamm aus sichtbare Backsteinfassade des Kesselhauses wurde unter größtmöglicher Berücksichtigung der historischen Bauteile erhalten und wieder instand gesetzt. Zu berücksichtigen war hierbei, dass die Fassade dieses Gebäudes zuvor nicht von der Straße aus sichtbar war: Die historische Stadtzentrale war im Blockinneren errichtet worden und jahrzehntelang durch die bestehende Blockrandbebauung des repräsentativen Verwaltungsgebäudes am Schiffbauerdamm verdeckt. Lediglich die östlichen Seitenfassaden des Kesselhauses und des Werkstattgebäudes waren von der im Blockinneren liegenden internen Werksstraße erkennbar gewesen. Am Werkstattgebäude ist als einzige Reminiszenz an die

better sound-proofing. For extra darkness, if needed, electrically operated shutters and blinds were added. The flooring in the studio area of the boiler-house was given a layer of parquet to make it a double-floor surface. For the outside of the main entrance area, which was also to serve as an observation or security area, a floor-cover of natural stone was envisaged. Carpets were laid in the remaining areas of offices, technical rooms and hallways. The main entrance area in the boiler-house was given a porch as an entrance-gate, as well as a construction of steel and glass to separate it from the main studio area. The buildings' heating was provided by the Bewag long-distance heating supply already present on the premises. The building conversions for RTL were planned in outline very soon after the site purchase; and a start was made immediately after the contract had been signed on its conversion and on site-utilisation on a rental basis. The other existing buildings, especially the – also listed – historic Bewag administrative headquarters in the building on 35 Luisenstraße, were also adapted, as necessary, to the office premises of the German Bundestag.Thanks to long years of use by the Bewag, the front building was in a very fine state of preservation after reunification, so much so, that modifications to the construction to match the specific requirements of buildings near the Bundestag could be kept to a minimum.

Appraisal of the Renovation Plan

The renovation plan and its implementation raised sensitive issues about the historic components of the erstwhile boiler-house, now an industrial monument. The brickwork façade of the boiler-house, visible from Schiffbauerdamm, was retained and repaired, – after the maximum possible account had been taken of the historic sections. It was important to respect the fact that the façade of this building had previously not been visible from the street. The historic city headquarters had been built in the centre of the block; and for several decades, it had been hidden by the existing perimeter block-construction of the imposing administration building on Schiffbauerdamm. Only the eastern side-façades of the boiler-house and workshop buildings were recognisable from the internal works road, situated within the block. The sole

Ehemaliges Kraftwerk Schiffbauerdamm,
Gesamtansicht von der Spree, Aufnahme um 2000
Former Schiffbauerdamm Power Station,
global view from the Spree, photograph around 2000

historische technische Nutzung noch ablesbar, dass auf den noch in der Außenfassade erkennbaren auskragenden Auflagern die ursprüngliche Versuchsstrecke einer Hochbahn montiert war.
Die Umnutzung als Fernsehstudio erfolgte vor allem aufgrund der herausragenden Lage in direkter Nähe zum Regierungsviertel und zum Sitz des Deutschen Bundestages im Reichstag. Die freigestellte Lage mit einem nach Süden orientierten Vorplatz erhöhte die Attraktivität des Standorts zusätzlich.

Weitere Entwicklungschancen des Quartiers

Nach der Realisierung der Baumaßnahmen für die RTL-Studios und weitere redaktionsnahe Bereiche im viergeschossigen Werkstattgebäude und den ebenfalls umgesetzten Maßnahmen für die Unterbringung von Bundestagsbüros wurden keine weiteren Entwicklungen in dem Quartier südlich des S-Bahn-Viaduktes vorgenommen. Es ist davon auszugehen, dass der Bereich östlich der Luisenstraße bis hin zum Bahnhof Friedrichstraße lediglich in einer langfristigen Perspektive unter weiterer Berücksichtigung der Flächenentwicklung des Deutschen Bundestages zu betrachten ist. Ein Abschluss der städtebaulichen Entwicklungsmaßnahmen scheint daher erst nach dem Jahr 2010 realistisch.

surviving aspect of the workshop building that recalls its historical technical purpose are the projecting supports – still recogniseable on the outer façade – from which a trial section of an elevated railway was originally suspended. The conversion into a television studio came about mainly because of the site's superb location – a stone's throw away from the government district and the seat of the German parliament in the Reichstag. Furthermore, the open position with a forecourt facing south greatly enhanced the site's appeal.

Opportunities for the District's Further Development

After the completion of the building plans for the RTL studios and other areas near editorial facilities in the four-floor workshop building, as well as conversion measures to accommodate the Bundestag offices, no further redevelopment was undertaken in the district south of the S-bahn viaduct. It is likely, however, that the area east of Luisenstraße – as far as Friedrichstraße station – will be considered solely as part of a long-term plan to develop the site occupied by the German Bundestag. All in all, then, it is not realistic to suppose that the process of urban regeneration will cease until 2010.

Hilmar Bärthel
Jahrgang 1921; Studium zum Vermessungsingenieur 1951 in Köthen, zum Diplom-Ingenieurökonom 1960 an der TH Dresden; Promotion 1975; tätig an der Deutschen Bauakademie in Berlin und im Büro für Tiefbauplanung und Koordinierung des Berliner Magistrats; Lehrbeauftragter an der TU Dresden und der Kunsthochschule Berlin-Weißensee; seit 1990 im Ruhestand.

Klaus Bürgel
Jahrgang 1931; Studium der Elektrotechnik an der Technischen Universität Berlin; 1983 Promotion; seit 1962 im „Elektrotechnischen Labor" der Bewag; ab 1972 Leiter der Netzplanung der Bewag; Lehrbeauftragter an der TFH- und TU-Berlin; seit 1993 im Ruhestand.

Matthias Dunger
Jahrgang 1959; Studium der Kunstgeschichte in Leipzig; 1990 Promotion; seit 1991 Denkmalpfleger im Landesdenkmalamt Berlin.

Hilmar Bärthel
Born in 1921; studied as a land surveyor in Köthen in 1951, as a qualified engineering economist in 1960 at the Technical College in Dresden; PhD in 1975; work for the German Academy for Construction in Berlin and in the office for civil engineering management and coordination of the Berlin Magistrates' Office; lecturing duties at Dresden Technical University and Higher Education College of Art, Berlin-Weißensee; retired since 1990.

Klaus Bürgel
Born in 1931; electro-technology studies at the Technical University in Berlin; PhD in 1983; from 1962, with the Bewag "Electro-Technical Laboratory"; from 1972, planning director of the Bewag grid; lecturing duties at Berlin Technical Higher Education College and Technical University; retired since 1993.

Matthias Dunger
Born in 1959; studied art history in Leipzig; PhD in 1990; from 1991 onwards, specialist for listed buildings in the Regional Office for Listed Buildings, Berlin.

Hans Achim Grube
Jahrgang 1966; Studium der Architektur in Berlin, Darmstadt und Florenz; 2002 Promotion; Projektleiter im Architekturbüro Max Dudler; Referent im Bundesbauministerium; seit 1996 Leiter der Bauabteilung und Leiter des Bereiches Immobilien und Allgemeine Dienste der Bewag AG, zusätzlich Geschäftsführer der Bewag Immobilienmanagement GmbH und Leiter Immobilien Vattenfall Europe AG.

Jörg Haspel
Jahrgang 1953; Studium der Architektur und Stadtplanung in Stuttgart; Studium der Kunstgeschichte und Kulturwissenschaft in Tübingen, Promotion; Oberkustos am Denkmalschutzamt Hamburg; seit 1992 Landeskonservator und Leiter des Landesdenkmalamtes Berlin.

Christina Keseberg
Jahrgang 1950; Studium der Architektur an der Technischen Universität Karlsruhe, Diplom; seit 1976 als Architekt für Planung, Bau- und Projektleitung in Berlin tätig; ab 1983 Architekt und Projektleiter bei der Bewag AG.

Hubert Staroste
Jahrgang 1952; Studium der Wirtschaftsgeschichte in Berlin; 1984 Promotion; Leiter des Denkmalarchivs im Märkischen Museum Berlin; Mitarbeiter des Instituts für Denkmalpflege; seit 1991 Mitarbeiter des Landeskonservators und des Landesdenkmalamtes Berlin.

Herbert Strobel
Jahrgang 1953; Ausbildung bei der Bewag zum Starkstromelektriker; ab 1990 Mitglied des Betriebsrates, 2000 Vorsitzender des Betriebsrates; ab 1992 Mitglied des Aufsichtsrates für die Arbeitnehmervertreter; 2000 Mitglied des Präsidiums des Aufsichtsrates. Seit 2002 Mitglied des Vorstandes der Bewag.

Hans Achim Grube
Born in 1966; studied architecture in Berlin, Darmstadt and Florence; PhD in 2002; project leader at Max Dudler Architectural office; consultant in Federal Ministry of Construction; from 1996, director of Buildings department and director of Property and General Services of the Bewag AG, in addition, managing director of Bewag Real Estate Management GmbH and director Real Estate of Vattenfall Europe AG.

Jörg Haspel
Born in 1953; studied architecture and urban planning in Stuttgart; art history and cultural science studies and PhD in Tübingen; head custodian in Hamburg Office for Listed Buildings; from 1992, regional curator and director of Regional Office for Listed Buildings in Berlin.

Christina Keseberg
Born in 1950; studied architecture at the Technical University, Karlsruhe, diploma; from 1976, work as an architect for planning, construction and project management in Berlin; from 1983, architect and project manager at the Bewag AG.

Hubert Staroste
Born in 1952; studied the history of economy in Berlin; PhD in 1984; director of Archive for Listed Buildings in the Berlin Märkisches Museum; associate employee of the Institute for Listed Building Preservation; since 1991, associate employee of Regional Curators' and Regional Office for Listed Buildings in Berlin.

Herbert Strobel
Born in 1953; qualified as an electrician for heavy current at the Bewag; from 1990, committee member of works board, in 2000, chairman of works board; from 1992, member of board of directors as employees' representative; in 2000, presidential member of board of directors. From 2002, Bewag directors' member.

Arndt, Hans Joachim: *Zusammenfassendes Gutachten Kraftwerk Rummelsburg*, Bewag-Archiv, Berlin 1995.
Bärthel, Hilmar: *Aufbruch Berlins zur „Elektropolis" – eine Zwischenbilanz im Jahr 2003* (unveröffentlichtes Manuskript), Bewag-Archiv, Berlin 2003. Die Texte in der vorliegenden Publikation sind Auszüge aus diesem Manuskript.
Bärthel, Hilmar: *Historische Kraftwerke in Berlin* (unveröffentlichtes Manuskript), Bewag-Archiv, Berlin 2002.
Berlin und seine Bauten, Teil I, Berlin 1896.
Berlin und seine Bauten, Teil XI: Industriebauten, Bürohäuser, Berlin 1971.
Bewag: *50 Jahre Berliner Elektrizitätswerke 1884-1934*, Berlin 1934.
Bewag: *Architektur und Kunst – Kraftwerk Moabit*, Berlin 1990.
Bewag: *Elektropolis Berlin, Historische Bauten der Stromverteilung*, Berlin 2000.
Bewag: Geschäftsberichte der Bewag und ihrer Vorgänger 1904-2001.
Bewag: *Hundert Jahre Strom für Berlin*, Berlin 1984.
Das Großkraftwerk Klingenberg, Charlottenburg 1928.
Grube, Hans Achim: *Renaissance der E-Werke, Historische Industriearchitektur im Wandel*, Berlin 2003.
Ingenieurwerke in und bei Berlin, Berlin 1906.
Kahlfeldt, Paul: *Hans Heinrich Müller*, 1879-1951 – *Berliner Industriebauten*, Basel 1992.
Matschoß, Carl: *50 Jahre Berliner Elektrizitätswerke*, Berlin o.J. (1934).
Neumeyer, Fritz: *Die Entwicklungsgeschichte der Berliner Stromversorgung / Das Kraftwerk Charlottenburg / Das Kraftwerk Moabit* (unveröffentlichte Manuskripte), Landesdenkmalamt Berlin, Berlin 1984.
Planungskontor für Städtebau und Ortsentwicklung und ProStadt: *Städtebauliche-landschaftsplanerische Rahmenplanung, Nalepastraße 10-50*, Berlin 2000.
Posener, Julius: *Berlin auf dem Weg zu einer neuen Architektur*, München 1979.
Spath, Christian und Nagel, Thomas: *Die Bauwerke der Berliner Stromversorgung* (unveröffentlichtes Gutachten), Landesdenkmalamt Berlin, Berlin 1988.
Wiese, Jens: *Kraftwerk Rummelsburg* (Diplomarbeit), Bewag-Archiv, Berlin 2002.

Bildquellen Picture Credits

Bärthel, Hilmar (Übersichtspläne und Entwicklungsstufen der Kraftwerke): S. 48/49 oben, S. 62/63 oben, S. 76/77 oben, S. 88/89 oben.
Bewag, 50 Jahre Berliner Elektrizitätswerke 1884-1934: S. 10, S. 11, S. 102 unten.
Bewag, Die Berliner Elektrizitätswerke bis Ende 1896: S. 102 oben.
Bürgel, Klaus und Bewag-Netzplanung (4 Schemakarten, Kraftwerke in Berlin 1907-2002): S. 14/15 oben.
Dipl.-Ing. Architekten Christina und Knud Petersen: S. 80, S. 81 oben rechts und oben links, S. 82 unten links, S. 84 unten.
Görner, Reinhard: S. 93.
Grube, Hans-Achim: S. 94 unten, S. 95 unten, S. 108.
HKW Klingenberg, Archiv: S. 30 links.
J.S.K. - Dipl.-Ing. Architekten: S. 104.
Kahlfeldt, Paul: S. 82 oben, S. 88 unten, S. 90 unten links, S. 92, S. 94 oben.
Landesdenkmalamt Berlin, Archiv: S. 7, S. 9, S. 38 oben, S. 99, S. 100, S. 101.
Pinardi, Mara: S. 97.
Profitlich, Florian: Umschlag.
RTL: S. 106 links, S. 107.
Schraps, Stephan: S. 32 oben links.
VDE-Sonderdruck „Bewag", 1993: S. 15 unten rechts, S. 22, S. 23, S. 28.

Alle übrigen Abbildungen stammen aus dem Bewag-Firmenarchiv.

Herausgeber Editor
Bewag Aktiengesellschaft
Puschkinallee 52
12435 Berlin

Konzept und Redaktion Concept and Editing
Hans Achim Grube
Christina Keseberg

Übersetzung Translation
SWB Communications: Suzanne Kirkbright und Anthony Vivis

Gestaltung Design
Susanne Rösler

Lithografie Lithography
Galrev Druck- und Verlagsgesellschaft, Berlin

Druck und Bindung Printing and Binding
Kästner Druck, Berlin

jovis Verlag GmbH
Kurfürstenstraße 15/16
10785 Berlin

www.jovis.de

ISBN 3-936314-01-2